The Evolving Coast

THE EVOLVING COAST

Richard A. Davis, Jr.

SCIENTIFIC AMERICAN LIBRARY

A division of HPHLP
New York

Library of Congress Cataloging-in-Publication Data

Davis, Richard A. (Richard Albert), 1937–
   The evolving coast/Richard A. Davis, Jr.
     p.   cm.
   Includes bibliographical references and index.
   ISBN 0-7167-5042-2
   ISBN 0-7167-6021-5 (pbk)
   1. Coast changes. I. Title.
GB451.2.D39 1993
551.4'57—dc20

93-25452
CIP

ISSN 1040-3213

Printed in the United States of America.

Scientific American Library
A division of HPHLP
New York

Distributed by W. H. Freeman and Company
41 Madison Avenue, New York, NY 10010
Houndmills, Basingstoke RG21 6XS, England

First printing 1996, RRD

This book is number 48 of a series.

To Mary Ann, who has been with me on almost every coast and is always ready to visit another one.

Contents

Preface

The surface of the Earth is characterized by an incredible variety of terrain: from mountains to valleys, from plains to plateaus. The geological formations that support these features vary widely in their age and composition, but most are hidden below layers of sediment. It is at the coast that we can see clearly the interaction of natural physical processes with the underlying geological elements.

The shoreline offers the observer the chance to watch beach materials being reshaped in only seconds as waves expend their energy. What an observer cannot grasp as readily, however, is the overall picture of how this complex segment of the Earth's surface evolves over long periods of time. It has been one of my aims in writing this book to show how the essential nature of the coast is determined by the large-scale character of the Earth's upper layer, especially the movements of this layer as described by the theory of plate tectonics. The tectonic reshaping of the Earth's surface interacts with fluctuations in sea level, both global and local, to create coastal scenarios on a scale thousands of kilometers in extent and thousands to millions of years in duration. More local and rapid changes are occasioned by physical processes such as weather, waves, and tides. The action of these processes on the coastal environments produces changes that extend only a few meters to many kilometers and that can occur in a matter of seconds, hours, or days. Each of these processes, from the large scale to the small scale, has its own origins and consequences, explored in the first three chapters of this book.

The challenge I faced in the remaining chapters was to show how varying geological settings and physical processes create coastal environments of many different kinds. The most important among these environments, and those I have explored at greatest length, are estuaries, deltas, barrier island complexes, and rocky coasts, for most of the world's coasts comprise combinations of these settings.

In writing this book, I have tried to provide the reader with an overall understanding of the forces that created our seacoasts and the depositional and erosional processes that cause them to continue to evolve, in both space and time. In focusing on the geological foundations of the coastal environment, I have purposefully neglected biological formations, such as coral reefs, and, to a lesser extent, biological systems that have an impact on coastal change. An adequate treatment of these topics would have demanded more space than this volume could provide.

As more and more of us live and vacation along the coast, this fragile portion of the earth is becoming threatened both by overuse and by misuse. The coast is not indestructible, but its destruction will not come via the waves and currents, but from human intervention in the natural systems. A better understanding of the evolution of coasts and their dynamics will enable us all to make informed decisions about their future.

I am grateful to Jerry Lyons, with whom I have had an editorial relationship for 25 years, for his confidence in this project and for his advice. The team of editors at W. H. Freeman and Company especially Mary Shuford, Susan Moran, Diane Maass, and Fay Webern have been the most skilled and patient that I have known. Travis Amos, the photo editor, gave me a new perspective on coastal photography as he searched far and wide for excellent photos. George Kelvin, the artist, did terrific work to bring many of the line illustrations to life. The first draft was written during sabbatical leave from the University of South Florida while in residence at the University of Waikato, New Zealand, and the Universities of Melbourne and Sydney, Australia. All these institutions are thanked for their hospitality and support.

Richard A. Davis, Jr.
Tampa, Florida
August 1993

Since *The Evolving Coast* was first published, I've been gratified that a number of instructors have used it in college and university courses, though the book was obviously not written as a textbook. In paperback, the book will now be more affordable to students and to other readers. For this version, I have corrected minor errors and added a list of readings for each chapter to help those who would like to explore topics in more depth.

I hope that you enjoy the book. Your comments are always welcome.

Richard A. Davis, Jr.
Tampa, Florida
October 1996

The Evolving Coast

Prologue

Along the Pacific coast of Washington and Oregon, the Earth's crust has been folded and faulted to create the Coast Range, whose slopes reach the ocean as irregular cliffs. For millions of years, the ocean surface has intermittently washed against this rugged mountainous area. The intersection of sea and land has, however, been at its present location for a few thousand years, at most. In these few thousand years, waves and currents have worn away pockets in the softer rock of the cliffs. Some of the eroded sediment has been carried to the deeper waters of the adjacent continental shelf; some is in solution in the sea; and a portion has been deposited in the eroded pockets in the form of beaches. The action of the sea has thus transformed the Pacific shoreline into a series of bedrock headlands interrupted by beaches.

Throughout the world, coasts are zones of transition between the ocean and land, where waves, currents, and tides act to mold the landforms, which in turn influence the movement of the water. From the meeting of sea and land at the shore, the coast extends inland until it leaves the reach of tides and storm waves and seaward until the bottom loses contact with the mo-

The rugged, eroding sandstone coastline at Cape Kiwanda, Oregon with large waves typical of the Pacific coast of the United States.

tion of the waves. The width of a coast can be less than 1 kilometer on cliffed coasts or more than 100 kilometers in large estuaries such as the Delaware Bay.

Above sea level, the action of rain, rivers, wind, and glaciers breaks rock into small particles, which are carried to the sea and deposited to form accumulations of marine sediments. Like all coasts, the Washington coast resembles dry land in its vulnerability to erosion, yet, like the ocean floor, it is the site of sediment accumulation. The processes of erosion and deposition vie to create landforms of great variety. Where erosion is dominant, the coast is rugged and craggy, as in Washington; where deposition prevails, as on the east coasts of North and South America, wide coastal zones stretch to the sea, marked perhaps by wide beaches and high dunes, deltas, and barrier islands.

The energy of the waves is able to transform the coastline noticeably in a brief span of time; in geological terms even the thousands of years required to give the Washington coast its present shape is short. Yet coastal processes can act faster still, in years or even days. Winter storms can

Sea oats growing on low dunes along a low wave energy, depositional coast, such as is common along much of the Atlantic coast of the United States.

2

destroy a beach within a day or two, and swift tidal currents can create a shallow channel in months. Coasts are one of the fastest changing parts of the Earth's surface, and they offer scientists and laypeople alike the chance to observe the progress of their geological development in only a few years.

Although waves can act quickly, especially on soft sediments, they expend their energy on a particular coastal setting that has taken millions of years to evolve. The overall configuration of the coastal areas of the Earth's crust is the product of as much as 180 million years of development, the period since the breakup of the continents to form the proto-Atlantic Ocean. Some coasts have been evolving for all that time; the settings of others have developed more recently but are still a few million years in age.

Throughout geological time, the sea level has been slowly changing relative to the coast, falling for thousands of years, then rising for thousands more, in response to global climate changes. During the few thousand years that the present Washington coast has been forming, for example, sea level has been rising about 1 to 2 mm a year. As the position of the shoreline changes, the coastal processes and their effects are translated across the shallow continental shelf and the adjacent coastal zone. The result is long and slow, but steady, change.

Any single coast is the result of processes at all three time scales: the slow geological processes of mountain formation and erosion that require millions of years; the gradual sea level changes requiring thousands of years; and superimposed over these the day-to-day and year-to-year combination of long-term and short-term action of the waves, currents, and tides. This book shows how the combination of long-term and short-term processes produces the deltas, estuaries, and barrier islands created by deposition and the rocky coastlines formed where erosion dominates.

Scientists have only recently undertaken comprehensive investigations of the coast. Although there was some interest shown during the nineteenth century in the origins of barrier islands, coral reef coasts, and the cliffed coasts of Brittany and the British Isles, the first systematic studies of the coast were conducted by geomorphologists in the early twentieth century. These scientists, who investigate landforms, had considered mountains, deserts, rivers, and glaciers for many years and finally began to view the coasts in the same way. They produced various classifications, maps, and studies of particular coastal features, especially along the northeastern part of the United States. Some of these scientists began to think about how coasts evolve and the processes responsible for this evolution. For example, Douglas W. Johnson wrote a classic work titled *Shore Processes and Shoreline Development* in 1919.

Also interested in the coast were the engineers who on the one hand constructed harbors, docks, and bridges, and on the other worked to stabilize and protect the open coast. Dikes have been constructed for centuries along the North Sea coast of Holland and Germany, both for protection and for land reclamation. In most other coastal areas engineers constructed structures that were designed to prevent or at least slow erosion.

Until World War II, these activities represented the focus of scientific and technological efforts to understand and, in some respects, to control the coasts. When the war came, the military began landing troops, supplies, and equipment on coasts throughout the world. Consequently, all branches of the military became interested in studying coastal geomorphology and coastal processes, including waves, tides, and currents, and in analyzing weather patterns along the coast. Much of the coast throughout the world was mapped in detail during this period. The Beach Erosion Board, a branch of the U.S. Army affiliated with the Corps of Engineers, conducted extensive research on beaches, waves, erosion, and other important aspects of the coast using both their own staff and scientists from many of the best universities.

Allied troops landing at Omaha Beach during the D-Day invasion of Europe in 1944; an excellent example of the importance of the coast in military operations. Such operations require special coastal characteristics, wave conditions, and tidal stages for landing massive numbers of troops and equipment.

Studies of this type continued after the war, but took on an engineering emphasis, a shift symbolized by the renaming of the Beach Erosion Board as the Coastal Engineering Research Center. During this same period the Office of Naval Research became heavily involved in basic research on the coast, initially through the Coastal Studies Institute of Louisiana State University. The institute directed its early efforts toward a global study of river deltas, beginning at home with the Mississippi delta. Later the Office of Naval Research expanded its support to include research into broad aspects of the world's coasts, especially beaches, inlets, and deltas—all potential settings for military activities. During the 1960s and 1970s this agency, under the leadership of Dr. Evelyn Pruitt, was responsible for much of the research on modern open coastal environments. A team of about two dozen scientists, primarily from universities, began what is practically speaking the modern era of coastal research.

Before the mid-twentieth century, then, coastal research consisted of simple observations and descriptions of coastal features. Only fairly recently have studies included the origin and development of these features through the processes that control the coast: its morphodynamics. It is this approach to an understanding of coastal evolution that this book will follow in its tour through the various coastal environments that typify the margins of the continents.

1

Plate Tectonics and the Coast

Coasts exist in a marvelous variety of forms. In the United States, for example, the Atlantic shore south of New England is bordered by broad, low plains washed by gentle waves. These coastal plains break up into a series of complex bays that are commonly fronted by barrier islands. In contrast, the narrow, rugged coast of the Pacific Ocean is marked by sculpted cliffs, pocket beaches, and crashing waves.

Whether the waves furiously pound or gently lap the shore, their constant action suggests a direct relationship between the sea and the shape of the coast. If we expand our view and examine coasts from a global perspective, however, the question arises as to whether there are less obvious, but more fundamental, reasons for the similarities and differences among the coasts. Is there a global-scale organization of the Earth's surface—a grand scheme—that controls the general nature of the coasts? This question has only been raised over the past few decades, since the development of the theory of plate tectonics. This comprehensive theory, one of the major scientific achievements of the twentieth century, explains the underlying forces that have created the mountain ranges, ocean trenches, and other

The San Andreas fault zone, which marks the boundary between the Pacific and the North American plates, emerges at the coast at Tomales Bay, California. Here the difference in the character of the landforms on the two plates is readily apparent.

major geological features of the Earth's surface. This new way of looking at the Earth both stimulated questions about the global organization of coasts and provided the means to answer them.

The Changing Map of the Earth

The theory of plate tectonics has a complicated history that reaches back to the global maps created after the great ocean voyages of the sixteenth and seventeenth centuries. As these maps became more accurate, the land-masses took on the appearance of pieces of a giant puzzle. Sir Francis Bacon is credited as the first to note this resemblance; in 1620 he wrote that the coastlines of South America and Africa would fit together perfectly if the ocean were not between them. Some scholars of the time concluded that the world's landmass had somehow split and separated in a series of catastrophes. Although the puzzling face of the Earth was the topic of passionate debates for nearly 300 years, the first attempt at a comprehensive scheme to explain the distribution of the continental landmasses was presented in 1912 by Alfred Lothar Wegener to the scientific societies in

Alfred Wegener was only 30 years old when he first proposed his theory of continental drift in 1912, the year in which this photograph was taken.

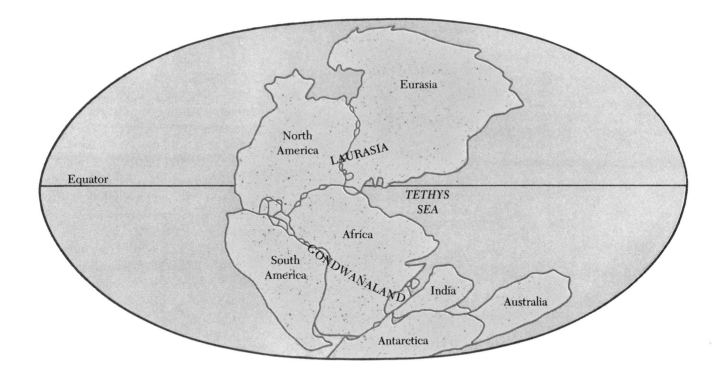

Frankfurt and Marburg. He believed that the continents had slowly drifted apart from a primordial supercontinent, which he called Pangaea (Greek for "all earth"). He envisioned a single world ocean, Panthalassa ("all ocean"), with a shallow sea, Tethys (from Greek mythology, the mother of all oceans), located between Laurasia and Gondwanaland, the northern and southern portions of the supercontinent. Using accepted geologic and paleontologic data, Wegener provided good supporting evidence for the continuity of geologic features across the now widely separated continents.

Three years later, Wegener produced his major work, *Die Entstehung der Kontinente und Ozeane (The Origin of Continents and Oceans)*, in which he presented an enormous amount of evidence in support of his theory. The book included examples of many types of geologic features that Wegener mapped to show continuity between the now-separated landmasses: an ancient mountain fold belt that seemed to continue from Canada to Greenland and on to Norway via Scotland, with identical sedimentary sequences appearing on opposite sides of the Atlantic Ocean; similarities in flora and fauna in different parts of the world; and evidence

According to Wegener, during the early Triassic Period (about 200 million years ago), all the continental land masses formed a single supercontinent, Pangaea. The northern portion of this supercontinent was called Laurasia, and the southern portion was Gondwanaland; between them was the Tethys Sea.

of glacial scars across Australia, South America, Africa, and India from the same icecap that included Antarctica. This evidence led Wegener to deduce that India was originally next to Antarctica but had become detached and drifted northward, where it collided with Asia to form the Himalayas.

Wegener's book was immensely popular. However, as his ideas gained attention and influence, they drew increasing ridicule from many leading geologists of the day, who were scornful of any suggestion that the massive continents were not fixed in the underlying rock. And indeed the weak link in Wegener's argument was his proposed mechanism for the movement of the continents. He suggested that the gravitational pull of the Moon and the Sun—the tidal force—exerts an uneven influence on the lithosphere, the Earth's rigid outer shell, that has been exacerbated over time by the spinning of the Earth. The land areas that experienced the stronger tidal force were pulled apart from the rest of the supercontinent. The continental crust, granitic in composition, is less dense than the basaltic oceanic crust. So as the supercontinent gradually broke up, the continental crust was able to drift over the underlying material.

Members of the general scientific community overwhelmingly agreed that the tidal force was entirely too weak to break up the continents and set them adrift. The repudiation of this part of his proposal made it easier for Wegener's critics to dismiss the rest of his theory. Wegener, however, was not discouraged by his critics. He devoted himself to gathering evidence for his supercontinent Pangaea until, on a fourth expedition to Greenland in 1930, at the age of 50, he disappeared while on a rescue mission.

After his death, a few loyal supporters of Wegener's theory continued the search for a drift mechanism. Unfortunately, some of their ideas were rather outlandish. For example, Howard Baker, an American paleontologist, suggested that a celestial body had passed close to the Earth and that the strong gravitational forces had pulled a piece of the crust away. The scar of this wound created the Pacific Ocean, and the excised crust formed the Moon. Baker theorized that the imbalance created by this event caused the remaining continental crust to break up and the pieces to drift apart.

The main centers of the scientific world reacted to these ideas by dropping all further discussion of the continental drift theory. For three decades after Wegener's death, it received little support except from a few academic centers in the Southern Hemisphere such as the University of Capetown in South Africa and the University of Tasmania in Australia. Prominent geologists at these institutions maintained their support for the idea and hosted various global conferences on the topic. A major symposium at the University of Tasmania in Hobart in 1956 led to a volume of

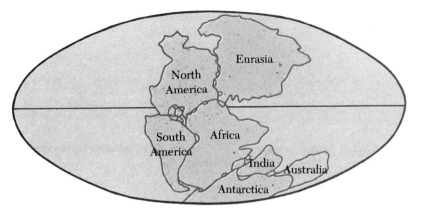

200 million years B.P.

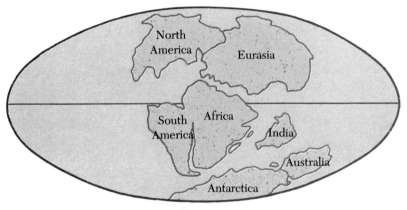

135 million years B.P.

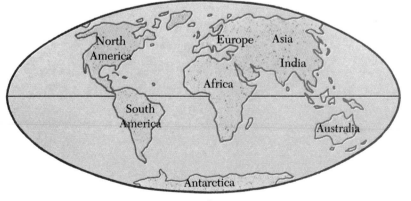

Today

The continental land masses that formed Pangaea gradually drifted from their original positions (top panel). They reached intermediate locations 136 million years ago, between the Jurassic and Cretaceous Periods (middle panel). After almost 200 million years, the continents reached their present positions (bottom panel).

collected papers on continental drift research from investigators around the world.

EVIDENCE FOR CONTINENTAL DRIFT

Early efforts to document the phenomenon of continental drift were seriously hampered because only the exposed landmass, 29 percent of the Earth's surface, was accessible for study. But if the shape of the oceans had changed as the continents separated, there should be some record of this in the ocean basins. What evidence should be looked for, and how could it be obtained?

Before World War II, very little was known about the ocean basins—not even the contours of the seafloor. But naval operations during the war,

The oceanic ridge system, which extends through the Indian, Pacific, and Atlantic oceans, outlines diverging plate boundaries. The oceanic trenches, which lie closer to continental edges, coincide with converging plate boundaries.

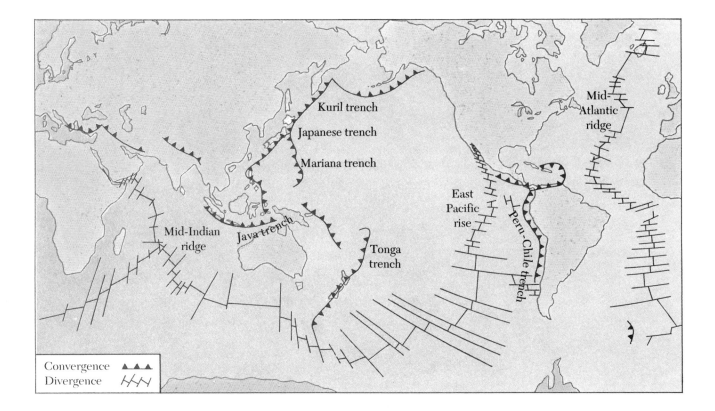

This underwater photograph taken on the ocean floor at the crest of the East Pacific rise shows a fracture in the Earth's crust at a zone of plate divergence. The white organism near the middle of the photograph is a deep-sea shrimp.

especially submarine warfare, created a demand for precise oceanographic information. Many senior geologists, physicists, chemists, and meteorologists were involved in gathering this information, either as naval officers or as consultants. After the war, these scientists were eager to assess the data they had collected and to continue their research. Thus, in the 1950s, with the aid of generous governmental support, especially surplus ships and equipment, serious study of all the world's oceans began.

Soundings of the ocean floor were a large part of the initial effort in this marine geologic research. One of the first major discoveries, by the late Maurice Ewing of Columbia University, was the presence of a virtually continuous mountain system that extends for 60,000 kilometers (km)—mostly under the oceans—around the globe. This mountain range, which is known as the oceanic ridge system, bisects the Atlantic Ocean and extends through the Indian and Pacific oceans and across part of Africa. Some of the peaks of this ridge system rise from the ocean floor to hundreds of meters above the ocean surface. Valleys extend lengthwise along the crest of this system and large fractures cut across it at numerous places along its course. These fractures are transform faults, places where the ridge system has been dislocated by crustal movement.

Paleomagnetic data from the minerals in the ocean strata have provided important clues about the dynamics of the Earth's crust. When minerals cool, their magnetic fields become oriented with the Earth's magnetic poles. Because the north and south magnetic poles switch once or twice every million years coincident with the reversal of the Earth's magnetic field, a record is left of each reversal in a series of stripes of opposite magnetic polarity in the oceanic crust. The patterns of the stripes on each side of the oceanic ridge are mirror images. Radiometric dating of rocks within these stripes shows that the age of the seafloor increases with distance on both sides of the oceanic ridge.

Other methods of dating oceanic materials also show a pattern of increasing age as one moves out from the oceanic ridges. The ages of the oceanic islands, known from dating the underlying rock, illustrate this phenomenon. Islands on or adjacent to the ridge—for example, Iceland—are quite young. Bermuda, which is more or less midway between the ridge and North America, is about three times as old as Iceland.

Sediments, composed in part of fragmented shells, accumulate gradually over the rocks of the ocean floor. These deep-sea sediments show an age pattern similar to that of the underlying rocks—the youngest sediments are closest to the ridge system. The oldest sediments under the western Atlantic Ocean lie off the New Jersey coast. In the Pacific, the oldest sediments are found near Japan, because the oceanic ridge system is

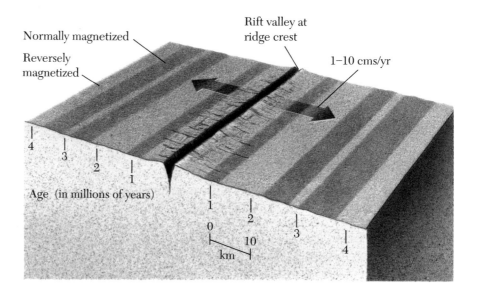

Magnetic stripes caused by alterations of normal and reverse polar magnetism in the oceanic crust occur in a symmetrical pattern on either side of the oceanic ridge.

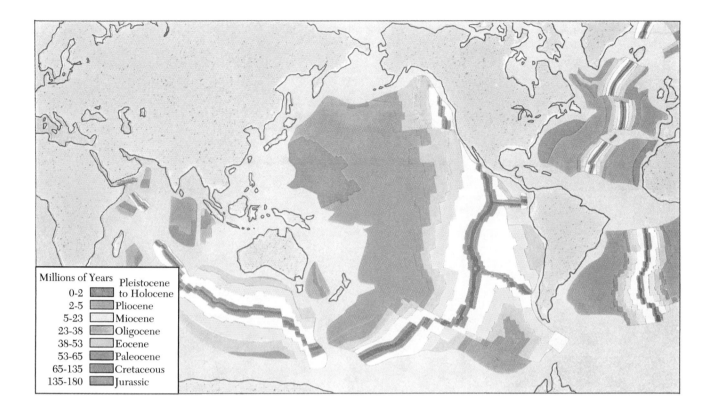

Millions of Years		
0-2		Pleistocene to Holocene
2-5		Pliocene
5-23		Miocene
23-38		Oligocene
38-53		Eocene
53-65		Paleocene
65-135		Cretaceous
135-180		Jurassic

close to South America. The oldest known sediments from any of the ocean basins were deposited during the Jurassic period; they are about 180 million years old.

The density of the oceanic crust also changes systematically—it is less in rocks near the mid-oceanic ridges and increases with distance away. The rock is evidently compacting over time; one result is that the ocean floor becomes gradually lower as it leaves the ridges.

The density and age patterns established two important facts: The stratigraphic record in the ocean basins is young relative to the age of the Earth; and there is a distinct pattern of increasing age away from the oceanic ridges. This pattern of increasing age suggested that the ocean's lithosphere was actually being created at the mid-oceanic ridges. The crust then moved away from the ridge on both sides, as it was pulled outward by the heavier, downward-trending older material. As the rock moved farther from the ridge, the ocean basin expanded and the continents moved farther apart. Here was the beginning of a theory that could explain the move-

Mapping the known ages of the oceanic crust demonstrates vividly that the geologically youngest areas lie nearest to the oceanic ridge system with a consistent increase in age farther from the ridges in both directions. The irregular distribution is caused by dislocations produced along the transform faults.

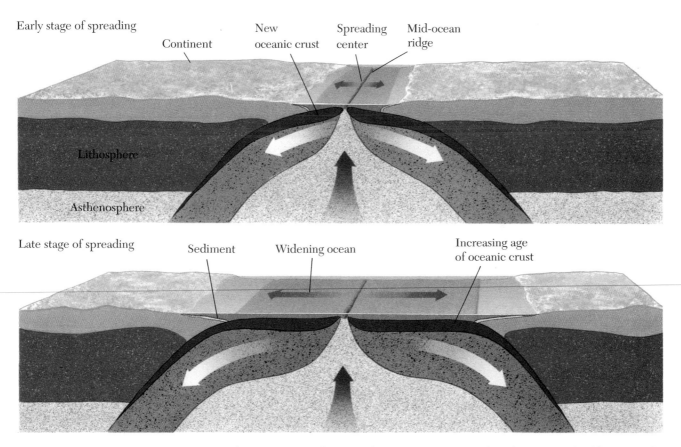

Early stage of spreading

Continent — New oceanic crust — Spreading center — Mid-ocean ridge

Lithosphere

Asthenosphere

Late stage of spreading

Sediment — Widening ocean — Increasing age of oceanic crust

A comparison of a spreading center associated with an oceanic ridge at early and late stages of separation shows the relationships between the major geologic elements as the ocean widens.

ments of the continents. The presence of the numerous transform faults across the ridges, sites of differential crustal movement across the ocean basin, also supported the notion that the ocean floor was spreading.

Heat flow measurements supported the emerging realization that dynamic processes shape the ocean floor. These measurements showed that the oceanic crust is warmest at the oceanic ridge, where it is the youngest. Because the rock forming the floor of the ocean basin is basaltic, the type formed by the magma or molten rock from volcanos, it seemed likely that the new crustal material was formed of hot magma welling from the ridge, which cooled and solidified as it aged.

Far from the mid-oceanic ridges, depth soundings revealed a series of deep trenches that parallel mountainous coasts or volcanic island arc chains. Although these narrow, elongate features are distributed throughout the oceans, they are most common in the Pacific. A plot of earthquake epicenters around the world reveals that the highest concentrations of epicenters occur at the trenches and their adjacent volcanic islands. Here earthquakes occur as much as 700 km (450 mi) below the surface. The second highest concentration of earthquakes is the oceanic ridge system; and there is high activity along the Himalaya Mountains and in the mountains of the eastern Mediterranean area.

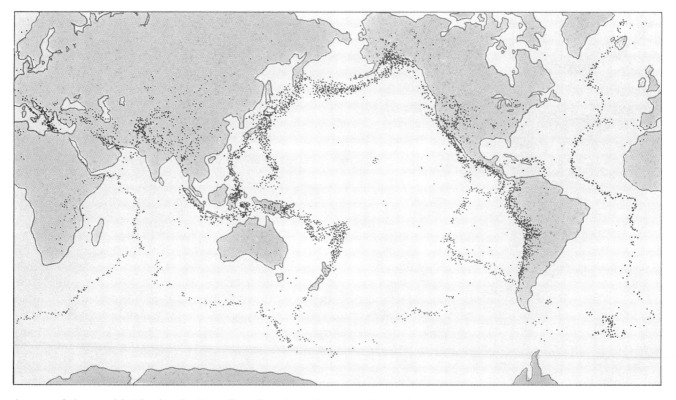

A map of the worldwide distribution of earthquake epicenters shows that their greatest concentration is associated with trench systems along the converging margins of plates and their next greatest concentration is along the oceanic ridge system.

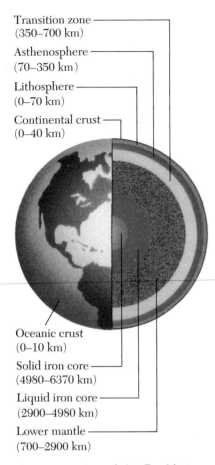

Transition zone
(350–700 km)

Asthenosphere
(70–350 km)

Lithosphere
(0–70 km)

Continental crust
(0–40 km)

Oceanic crust
(0–10 km)

Solid iron core
(4980–6370 km)

Liquid iron core
(2900–4980 km)

Lower mantle
(700–2900 km)

A cross-section of the Earth's interior shows the depth of its major layers.

Enough information was now at hand for a broad synthesis of the geological and geophysical findings and a new interpretation of the dynamics for the Earth's crust, called plate tectonics. Although we are still learning more about how it works, the theory is almost universally accepted because it explains almost all the broad-scale attributes of the Earth's crust.

PLATE TECTONIC THEORY

Almost as Wegener had proposed, plate tectonic theory states that the continents, as part of the lithosphere, the Earth's uppermost layer containing the crust, drift on the semimolten underlying material we call the asthenosphere, or the upper mantle. By the 1960s, scientists had concluded that the lithosphere is divided into 12 large, tightly fitting plates and several small ones. Six of the large plates bear the continents; the other six are oceanic. And, as Wegener asserted, all of the plates are in motion.

New lithosphere is continually being formed as basaltic magma emerges from the rift valleys of the oceanic ridge. These rift valleys are zones where the oceanic plates are diverging. Here, the crust is stretched and thinned enough to weaken it and allow magma to bubble up and build the oceanic ridge. As the new lithosphere cools, it becomes more dense and moves down the flanks of the ridge.

Lithosphere is also continuously disappearing. The island arc-oceanic trench systems are sites of plate convergence. Here, one lithospheric plate slips below another and descends back into the asthenosphere, where it is consumed. Its course can be traced by plotting the epicenters of the deep earthquakes at these zones, which release the stresses produced as the descending slab meets resistance from the underlying rock. Where the edges of an oceanic plate and a continental plate converge, it is the more dense oceanic plate that subducts into the asthenosphere. These convergence areas are called subduction zones. As the edge of the subducting plate descends, it forms a deep, asymmetrical oceanic trench.

As the descending plate moves into the asthenosphere, some of the rocks along its upper edge melt into magma. This magma rises back to the surface through faults in the plate margin, producing numerous volcanos that are arranged in a line parallel to the trench. Either an island arc or a chain of volcanos on the margin of the landmass is formed.

Oceanic plates thus constitute a kind of conveyor belt system: Lithosphere is formed at the mid-oceanic ridges, moves away toward the edges of the plate, and finally is subducted into the asthenosphere to become magma again. The movement is driven in part by gravity; the cooler, denser, older edge of the plate descends into the trenches and the warmer, less dense, younger material rises at the ridges.

Much of the motion of the lithospheric plates is a relatively simple spreading at the oceanic ridges and convergence at the trenches. Convergence of two continental plates produces folding and mountain building at the boundaries of both plates. However, these tectonic movements are complicated by the fact that the rigid lithospheric plates are moving over a spherical surface, so they are subjected to the strain exerted on hard, relatively flat objects forced to fit over a soft curved one. The numerous transform faults along the oceanic ridge system are places where some of this strain is relieved by fracturing and by small slip movements along these fractures. Stress is also relieved at the junctions where three plates converge. At these sites various combinations of divergent, convergent, and transform-fault conditions adjust the fit of the plates to the globe as they move.

The rate of plate movement varies from about 1 centimeter (cm) a year at the Mid-Atlantic ridge to 10 cm a year at the East Pacific rise in the southeastern Pacific. A few of the rates can be determined by direct observation from satellite data; the majority are calculated from the positions of marine sediments and magnetic minerals of known ages.

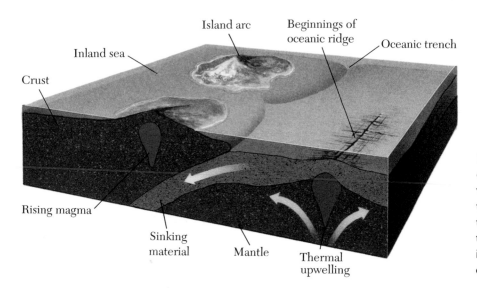

Magma rises through the oceanic crust both at the weakened rift valleys and along the oceanic trenches formed at plate subduction zones. Ridges are built up at the edge of diverging plates, and island arcs develop at the margins of converging plates.

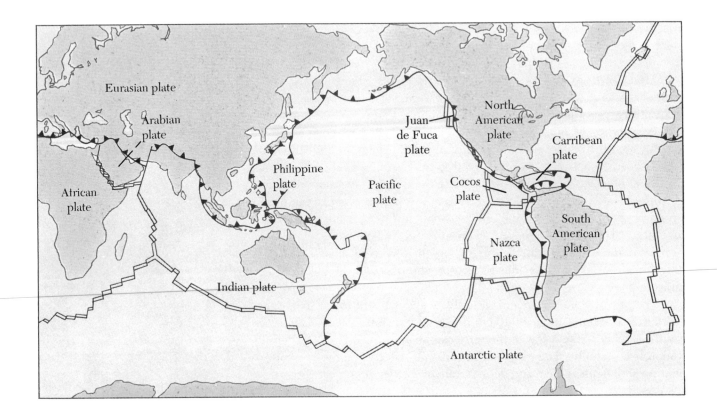

Eurasian plate

Arabian plate

African plate

Philippine plate

Juan de Fuca plate

North American plate

Carribean plate

Cocos plate

Pacific plate

Nazca plate

South American plate

Indian plate

Antarctic plate

The 12 major lithospheric plates are in constant motion. The outlines of their convergent and divergent boundaries make clear the directions in which each plate is moving.

Although his proposed mechanism was wrong, Wegener's theory of continental drift was proved correct. The continents separate, drift, and collide at a very slow rate; but they do so on the backs of tectonic plates. Plate tectonics, the study of plate boundaries and movements (*tectonics* is from the Greek word for carpentry), has incorporated continental drift theory. It has become the basis for explaining the processes producing the major features of the Earth's surface, including the coastlines of the continents.

THE TECTONIC CLASSIFICATION OF COASTS

Plate boundaries are regions of immense geological activity, where frequent earthquakes are symptoms of the rapid deformation of the Earth's crust. In contrast, the interiors of the plates are relatively stable. The prox-

PLATE TECTONICS AND THE COAST

imity of a coast to a plate boundary, and the type of boundary, will have a tremendous influence on the fate of that coast. The processes of subduction and collision help shape the broad features of a coast near converging plates, whereas the tectonic stability of a mid-plate region will equally influence the form of a coast lying far within a plate.

From the global view of the Earth's crust provided by the theory of plate tectonics, Douglas L. Inman of the Scripps Institution of Oceanography and his colleague C. E. Nordstrom have developed a comprehensive model of the interrelationships between plate tectonics and coastal types. In 1971 they published their synthesis, a utilitarian tectonic classification of coastal types that is still used today.

Inman and Nordstrom suggested that the broad shape of a coast depends on which of three tectonic settings is most closely associated with that coast. Their classification divides all the continental coasts into three major types: those associated with the leading edge of a crustal plate, those associated with the trailing edge of a plate (these coasts are most often in

This schematic drawing illustrates the tectonically active coasts that are associated with converging plates (leading edge coasts) and those that are located away from actual plate margins (trailing edge coasts).

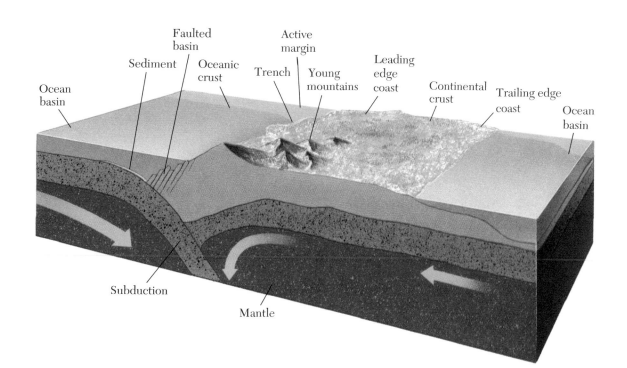

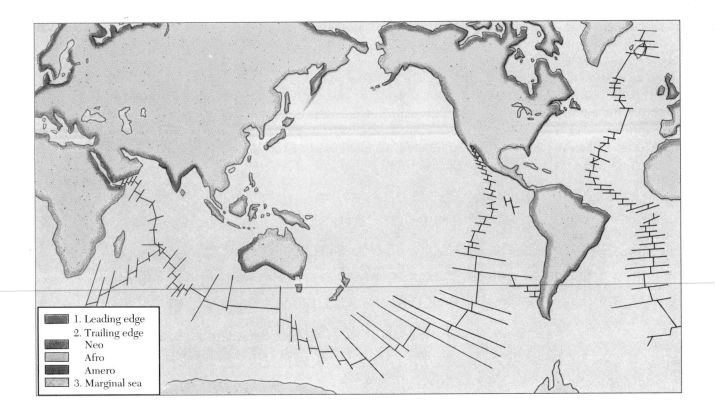

1. Leading edge	
2. Trailing edge	
Neo	
Afro	
Amero	
3. Marginal sea	

When mapped according to the Inman and Nordstrom classification system, the worldwide distribution of coastal types highlights the relationship between these various coastal types and the tectonic setting.

mid-plate), and those bordering a sea enclosed between the landmass and a volcanic island arc at the plate boundary. Island coasts are not considered in this classification.

Plate tectonics explains only the broadest features of the coast. The beauty of the Inman and Nordstrom scheme is its recognition that coasts contain features at several scales of size, molded by forces that operate at different scales. First-order features cover large geographic distances—a thousand kilometers or more. At this scale the broad character of the coast is directly tied to the major tectonic conditions of plate boundaries or intraplate settings. Second-order features are influenced by smaller-scale factors, for example, erosion and sediment deposition, which shape coasts over distances of tens to hundreds of kilometers. A good example of a second-order feature is the series of long, slender barrier islands forming the Outer Banks of North Carolina. Rock type and structural orientation become important on this scale. Third-order features are typically only a few kilometers in length. They are affected by short-term processes that last for years to decades and tend to be controlled by local conditions of deposition. Third-order features include beaches, which are caused by wave action; tidal inlets; and individual barrier islands.

Leading Edge Coasts

A leading edge coast develops along the border of a landmass where the oceanic edge of one plate converges with the continental edge of another. Also known as collision or convergent coasts, leading edge coasts are distinguished by rugged, cliffed shorelines regularly assaulted by large waves.

The convergence between the two plates may produce subduction zones as the denser oceanic plate descends beneath the continental edge of the other plate. The tremendous friction created by the converging plate edges causes the lighter continental crust to fold and buckle, creating the mountain ranges often set near leading edge coasts. In addition, rising magma may create volcanic ranges such as the Andes of South America and the Cascades of the northwestern United States. Because the angle of subduction is less steep under continental crust than under oceanic crust, the volcanic range may be some distance from the trench. Thus, the Cascades lie inland from the Coast Range.

The coast near Antofagasta, Chile, is a leading edge coast.

Where the oceanic plate is not subducting steeply, as off the coast of Washington and Oregon, considerable crustal material may be scraped from the plate and left behind to form a narrow continental shelf. In contrast, off the coast of South America, there is essentially no continental shelf, because the oceanic Nazca plate is descending so steeply into the Peru-Chile trench that no crustal material is scraped off.

The steep mountain slopes of leading edge coasts have rapidly flowing streams and small rivers that quickly erode their beds. Because the drainage divide is at a high elevation near the coast, the rivers are short, steep, and straight. They transport large quantities of sediments directly to the coastal areas, giving no opportunity for sediments to become entrapped in a meander, on a natural levee, or on a floodplain. The rivers dump their sediment loads into coastal bays or directly onto open beaches.

Once deposited, the sediment has no opportunity to accumulate to form a coastal plain, in part because of the narrow continental shelf. The rugged topography and high relief that characterize a leading edge coast of this type extend below sea level as well as above. The steep bottom gradi-

The schematic drawing shows the relationships between a stream, a beach, a littoral drift system, and a submarine canyon as they disperse sediment away from a leading edge coast.

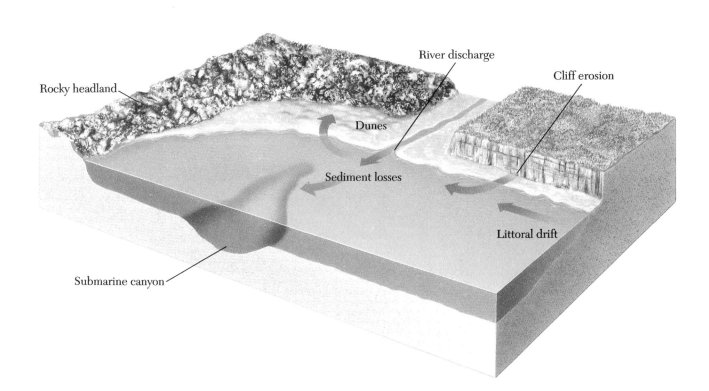

Rocky headland

River discharge

Cliff erosion

Dunes

Sediment losses

Littoral drift

Submarine canyon

A cascading sandfall travels down a submarine canyon in this underwater photograph.

ents and deep water near the coast permit the large waves of the Pacific to maintain their size as they approach land because there is no shallow sea bottom to interfere with wave motion and diminish the waves. The large waves strike the shore with tremendous energy, producing high rates of beach erosion along many parts of the Pacific coast.

These waves commonly approach and strike the coast at an angle, thereby causing sediment to be transported alongshore in the surf zone. The steep gradient on the California coast is punctuated with deep channels—called submarine canyons—that descend to depths of hundreds of meters close to the shoreline. As sediment is carried along the coast, it is intercepted by these submarine canyons and transported into deep water, where there is no mechanism for its return.

Even though mountain streams deposit large amounts of sediment on the coast, they do not produce deltas. In fact, none of the world's 25 largest deltas occurs on leading edge coasts, because this tectonic setting does not have a shallow, nearshore area on which the sediment can accumulate and what sediment does accumulate is soon dispersed by large waves.

Only small deltas like this one at the mouth of the Pisco River in Peru where it enters the Pacific Ocean, are likely to form near converging plate boundaries.

The only areas of significant sediment accumulation on collision coasts are bays, which form as the result of a variety of geologic circumstances. Some bays are created by flooding of stream valleys; others are caused by faulting and the resulting displacement of the crust. Many of the bays in the San Francisco area have developed as a result of its position on the San Andreas fault zone, where extensive crustal movement has occurred.

In California, the zone of plate contact is dominated by transform faults rather than a subduction zone. The most prominent of these faults is the well-known San Andreas fault system, the source of many of the earthquakes that plague this region. This system of faults connects a spreading ridge running through the Gulf of California with a spreading ridge off the coast of northern California. The ridges and the transform faults are part of the enormous boundary of the Pacific plate. In California, this plate is moving northwest along the San Andreas fault relative to the North American plate to the east.

Because of its position along the plate boundary, the southern California region—the land and the adjacent ocean—is a complex arrangement of geological features. Prominent among these are the numerous sedimentary basins that have formed as the result of transform fault tectonics. These

Faults

basins, termed "pull-apart basins" by John C. Crowell of the University of California, develop where a transform fault branches to form a wedge. As movement occurs along a sinuous fault branch, the wedge drops within the crust, forming a basin bounded by fairly high-angle faults on all sides that eventually fills part way with sediment. Dozens of these sedimentary basins exist in the southern California area both on land and under the ocean.

Many of these pull-apart basins began forming during the Miocene Epoch, about 10 million years ago. Since that time, the relative position of sea level in the area has changed and many of the older basins that long ago filled with ocean sediment are now above sea level. Some of their faulted margins form the irregular, hilly areas of the southern California coastal region. Those basins that are still below sea level are filling with sediment and forming the continental borderland—a complex of small, fault-bounded basins under the ocean.

Trailing Edge Coasts

The most diverse of the three main coastal types are those that form on the trailing edge (or passive) continental margins. Here the coast develops in association with a part of the continental lithosphere that is not at the edge

This coarse gravel beach along a high-relief coast on the Sea of Cortez, Mexico, provides an example of a Neo-trailing edge coast.

of a plate and that typically has been tectonically stable for at least tens of millions of years.

Inman and Nordstrom have categorized trailing edge coasts on the basis of their plate tectonic settings as Neo-trailing edge coasts, Afro-trailing edge coasts, and Amero-trailing edge coasts. The three subtypes represent a continuum of the conditions that develop as landmasses drift apart after breakup. Just after drifting begins, when spreading is initiated, the separating landmasses have coasts with high relief, small rivers, and little deposition—like leading edge coasts. As spreading continues at a rate of 1 to 2 cm per year, however, there is plenty of time for the coastal areas to erode. Their cliffs become low plains where sediment can be deposited, eventually to form deltas, barrier islands, and other sedimentary features. Once separated from the rift area, the margin remains tectonically stable until involved in another rift or a convergence of plates. During this entire process, the edge of the landmass changes from a Neo-trailing edge coast to an intermediate Afro-trailing edge type to an Amero-trailing edge margin, which is the most mature type because it contains the most extensive deposits of sediment and the widest continental shelf.

As just noted, a Neo-trailing edge coast occurs as plates diverge from an active spreading center where new oceanic crust is produced. Such a coast represents the first stage of coastal development and is only a few million years old. Coasts like this existed just after the proto-Atlantic developed, as the continents split up during the Triassic period, 190 million years ago.

The coasts of the Red Sea, where the Arabian and African plates are separating, and the Gulf of California, where the Pacific and North American plates are diverging, are good present-day examples of this type of coast. Both the Red Sea and the Gulf of California are long and narrow, representing very early stages of separation before a continental shelf and the rest of the continental margin have developed. Although steep nearshore areas are present, the narrowness of the water bodies prohibits large waves from developing. Therefore erosion along the coasts is minimal.

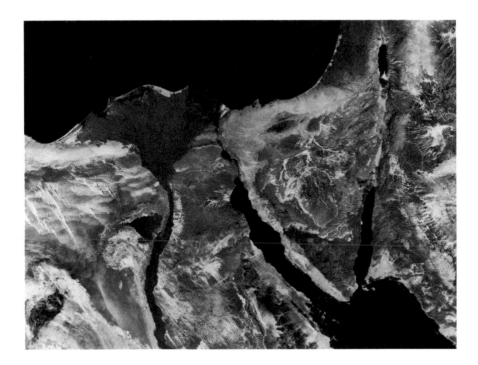

The Gulf of Suez on the left and the Gulf of Aqaba on the north end of the Red Sea mark diverging plate boundaries. The coasts of those bodies of water are examples of the Neo-trailing edge type.

Neither of these modern examples of Neo-trailing edge coasts is served by large river systems that supply sediments. However, this is due, in part, to their location in arid regions. A different situation was present when North America split from Africa in the Triassic Period—a time of warm and humid climates.

An Afro-trailing edge coast forms on a continent that has coasts of only the trailing edge variety. Such a continent occupies a position in the middle of a crustal plate that has little tectonic activity along its margins and has been relatively stable for many millions of years. The obvious example of this situation is the African continent, which is positioned near the center of the African plate and has no collision or convergence areas, except at the northernmost part in Morocco and Algeria along the Mediterranean Sea. Greenland, resting on the northeastern part of the North American plate, is another example.

Afro-trailing edge coasts have reached a stage of development in between those of the juvenile Neo-trailing edge and the mature Amero-trailing edge types. That is not to say that their absolute ages are necessarily in the same relative order. North America and Africa have remained in the same relative positions, with essentially no major tectonic interference,

Huge dunes of the Namibian desert meet the Atlantic Ocean along the coast of southwest Africa. This setting provides a good example of an Afro-trailing edge coast.

PLATE TECTONICS AND THE COAST

since their separation by the opening of the Mid-Atlantic ridge. Although both continents have trailing edge coasts of the same absolute age, the North American continental margin is more mature than that of Africa, as we shall see later.

Afro-trailing edge coasts have developed pronounced continental shelves and plains, but these features lack the extent of more mature coasts, and sedimentary features such as large deltas are rare. On these coasts the abundant sediment for coastal development is not available along most of their length. The African continent has been relatively stable for a long time, so no extensive, high mountain ranges have developed. The modest to large river systems drain areas of only modest relief, and their sediment load is correspondingly modest.

Climate and other factors have introduced variations in the degree of coastal development. Climate influences sediment deposition through its role in sediment production. Rainfall fills rivers and weathers rocks to sand and mud—the typical materials carried by rivers. Most of the world's large river systems are located in those temperate and tropical regions that receive abundant rainfall. Thus, a well-developed continental margin with a broad shelf appears along the southwestern coast of Africa where abundant sediment has built up the Niger delta, but a much smaller continental margin exists on the eastern and western coasts, where river systems are smaller.

Alongside plate tectonics, climate is thus the other important influence on first-order coastal features, through its effects on erosion and deposition. Strong winds and intense storms produce large waves that cause rapid erosion of the coastline. Such conditions would be expected in high latitudes. At extremely high latitudes, the northern portions of North America and Asia, all of Greenland, and the continent of Antarctica, have precipitation only in the form of ice and snow throughout most of the year. Most of the water is in the form of ice. Thus, little water is available for the development of rivers that would carry sediment to the coast. Likewise, large areas of Asia, Africa, and the southwestern United States have poorly developed river systems because they are deserts. Where little sediment is transported to the coast, there is very slow development of a depositional coast—even in the absence of energetic waves.

Amero-trailing coasts, the most geologically mature coastal areas, are represented by the east coasts of North and South America. Both are tectonically stable portions of the continents, well away from the plate boundary, and have been so located for at least several tens of millions of years. For example, the area south of New England between the Appalachian

This Arctic coastal landscape at Somerset Island, Canada, is characterized by its high cliffs with debris slopes. It is a good example of a high-latitude Afro-trailing edge coast.

mountains and the outermost edge of the continental shelf, more than a hundred kilometers away, has been tectonically stable since the continents separated in the Triassic period, nearly 190 million years ago. If we look at the overall profile across the continental portion of this part of the North American plate, we see a wide, gently sloping, generally featureless surface underlain by sedimentary strata. The shoreline is located somewhere near the middle of this several-hundred-kilometer-wide profile. That location places the Amero-trailing edge coast near the middle of a very wide and rather flat portion of the continental part of a crustal plate.

The combination of long-term tectonic stability, a temperate climate, and the development of a broad coastal plain has provided huge quantities of sediment to trailing edge coasts since shortly after the continents separated. During this time numerous large, meandering river systems have developed. For more than 150 million years, these rivers have been carrying sediment across a gentle incline. As they have deposited sediment at or near their mouths, they have created broad, low-relief coastal plains on the landward side and, on the seaward side, gently sloping continental shelves.

Extensive mangrove stands and tidal flats cover the low relief Amero-trailing edge coast near the mouth of the Amazon River in Brazil.

Wave action along Amero-trailing coasts is limited because the shallow water of the gently sloping inner continental shelf interferes with wave motion. Large mid-ocean waves lose energy as they progress across the shelf, and consequently do not inhibit deposition of sediment along the coast.

As a consequence of many millions of years of deposition, the eastern United States south of New England has a wide, stable continental margin, which supports a wide coastal zone that continues to receive sediment. At the present time, most of this sediment is being trapped in the numerous coastal bays and estuaries that have formed as the rising sea level has drowned the river mouths over the past several thousand years. Chesapeake Bay, Delaware Bay, and Pamlico Sound are estuaries of this type along the mid-Atlantic coast.

South America is an example of how a drainage divide on the leading edge area of the plate influences the coast thousands of kilometers to the east on the trailing edge side. In that continent, much of the drainage into the Atlantic Ocean originates in the high Andes Mountains, which rise along the leading edge western coast along the Pacific, and flows from there across most of the wide part of the continent to the eastern coast. A profile across the South American continent shows a narrow, steep gradient extending westward from the drainage divide only a few hundred kilometers to the western leading edge Pacific coast. The eastern portion of the profile consists of a gently sloping gradient of thousands of kilometers to the eastern trailing edge coast. This profile illustrates perfectly the contrast between the topography of a leading edge coast and a mature trailing edge coast.

The subcontinent of India has a somewhat similar topography. The runoff from the distant highlands of the Himalayas, where the subcontinent is colliding with Asia after moving across what is now the Indian Ocean basin, drains into the Ganges-Brahmaputra, the Indus, and the Godavari. These three large rivers discharge into the Indian Ocean. Again we can see how climate and tectonic setting influence coastal development. The Bay of Bengal, where two of these rivers reach the coast, is an area where monsoon conditions dominate the climate. During the monsoon season, rainfall is extremely high, intense storms are common, and tremendous quantities of sediment are carried to this coast. A huge delta has developed there.

Other Amero-trailing coasts are the northern coasts of Europe and Asia on the Eurasian plate and throughout most of Australia on the Indian-Australian plate. They also extend across the northern margin of North America. Because of the rigorous conditions there and the lack of research, however, little detail is known about this high-latitude region.

A review of the plates of the Earth's lithosphere shows that Amero-trailing coasts do not exist on opposite sides of any present-day continent, although there is no reason why these mature coasts could not theoretically do so. The continent whose coasts are closest to this condition is Africa, which has the less-developed Afro-trailing edge coasts on both sides of its landmass.

Marginal Sea Coasts

On some tectonic plates, continental coasts are nearest to the plate boundary where a collision is occurring, but are kept apart from its influence. In these places, a moderate-sized marginal sea separates a passive and tectonically stable continental margin from the volcanic island arc and deep-sea trench that mark the plate edge at a subduction zone. In the South China Sea, the Philippine Islands form the leading edge island arc that protects a marginal sea from the open Pacific waves. For the Gulf of Mexico, the volcanic areas of Central America and the adjacent Caribbean are the plate-boundary island arcs.

This view of the Yangtze delta shows the river as it enters the East China Sea. The light blue cloudlike areas are suspended sediment being discharged at the river mouth.

Although fairly close to the convergence zone, the marginal sea coast is far enough away to be unaffected by convergence tectonics—it behaves like a trailing edge coast. Well-developed rivers carry large quantities of sediment to the coast, where a broad and gently sloping continental shelf provides an ideal resting place for large quantities of land-derived sediment.

The restricted size of the marginal sea limits the size of waves that develop. In addition, the gentle slope and shallow waters of the continental shelves in these areas attenuate wave energy. Hence, the combination of relatively low-energy coastal conditions and sizable sediment loads allows the formation of large deltas and other coastal sedimentary deposits such as tidal flats, marshes, beaches, and dunes. The great rivers of southeastern Asia and the Gulf region of the United States, both areas of mild climate and abundant rainfall, have deposited their sediment loads on marginal sea coasts to create some of the most impressive deltas in the world.

APPLYING THE TECTONIC CLASSIFICATION

The Inman and Nordstrom tectonic classification of coasts is an internally consistent system that can be widely applied to coasts around the world. It should be remembered, however, that this classification is only intended to be applied on a large scale. Tectonic interpretations cannot be made on the basis of local coastal morphology. For example, the southern coast of Australia, with its narrow beaches and nearly vertical cliffs, seems to have the characteristics of a leading edge setting; but it is not even near a plate boundary. It merely looks like a leading edge coast because the high-energy waves of the Southern Ocean have heavily eroded the shoreline. In contrast, the southern coast of Alaska has extensive tidal flats, marshes, and barrier islands. It looks like a trailing edge coast, but it is a tectonically active area at the collision boundary of the North American and Pacific plates. Its character is due to the retreat of glaciers. As the glaciers in the region melted over the past several thousand years, they left behind tremendous volumes of sediment of all particle sizes along the coast. Even though this part of the Alaskan coast is subjected to intense wave energy, the glacial sediment, with its heavy load of gravel, has ensured a net sediment accumulation. The tectonic classification of the Australian and Alaskan coasts becomes clear only when each is considered as part of a larger stretch of coast at the first-order scale of thousands of kilometers.

At the regional and local scales, the second- and third-order features become noticeable. These features may be imposed on almost any tectonic setting. Unlike the first-order features, they cannot be explained by a single all-encompassing explanation. They are the creation of a combination of processes that may act independently of one another. Second-order features, for example, are determined by the eroding power of large, frequent waves and the depositional tendency of smaller waves; by the sediment movement effected by tides, rivers, and glaciers; by the types of rocks present and their structural attitude; by climate; and by the supply of sediment. They are even determined by the activities of biological organisms. Marshy coasts and mangrove swamps are distinct coastal types that may extend for hundreds of kilometers where a stable tectonic setting of low relief and low wave energy prevails, such as at the border of a marginal sea or on an open coast of weak waves.

As the Earth's lithosphere changes over geological time, the regional and local coastal types change as well. Thus, many different coastal settings and their specific environments are incorporated into the stratigraphic record. We can recognize evidence of coastal characteristics in rocks dating well back into Precambrian strata, which were deposited at least 1 billion years ago. From these ancient coastal sedimentary deposits, it is possible to decipher the tectonic setting in which they accumulated.

The processes act on regional or even smaller scales to create and modify the coastal features to be discussed in the remainder of the book. Yet one particularly influential factor transforming the coasts—the changing sea level—can operate over the entire globe, like plate tectonics, while helping to shape the coast at a regional level.

2

The Changing Sea Level

All the wave and current energy that is expended on coasts, eroding the shore and transporting sediments, is confined within a vertical zone of only 20 m: Storm waves and tides generally reach no more than 10 m above the mean sea level, whereas at 10 m below sea level waves have become too weak to transport sediments, except under storm conditions. Because the coastal processes act over so narrow a vertical range, a change in sea level can leave former coastal features submerged below sea level or raised out of the water's reach. An earthquake can uplift or sink a block of land by more than 15 m in only a moment. Less dramatic but more significant, the sea level of the entire globe has fluctuated repeatedly in the last 2 million years as a succession of ice ages trapped masses of ocean water in huge sheets of ice.

The lowering of the mean global temperature in an ice age freezes large volumes of water into polar icecaps and continental glaciers, lowering the global, or eustatic, sea level. When global temperatures increase again, the glaciers melt and discharge this water back into the oceans. During the course of these episodes, sea level may change by more than 100 m, enough to erode and

Pt. Long at San Diego, California, is a converging coast that has experienced considerable sea-level change over the past several thousand years.

build a whole series of coasts. As the surface of the sea moves away from well-developed coastal areas, the waves expend their energy on new landforms, perhaps of different rock type and history, initiating a new cycle of development. A complex history of emergence and submergence can be interpreted on almost any coast.

This chapter considers the tectonic and climatic factors that cause sea level changes and the gross effects of these changes on coasts. Although tides do affect coasts, the periodic rise and fall of the seas with the tides do not change the net local sea level. Therefore, a consideration of tides will be postponed until the next chapter in which coastal processes are considered.

TECTONIC ACTIVITY

Seismic activity along an unstable portion of the Earth's crust may cause a sudden shift in regional sea level by sinking or uplifting the shore. On March 27, 1964, a severe earthquake along the convergence boundary of the Pacific and North American plates brought spectacular changes to the southern coast of Alaska. The quake registered a magnitude of 8.6 on the Richter scale near the epicenter on the north shore of Prince William Sound between Anchorage and Valdez, making it one of the most severe

The coast area near San Juan, Peru, along the tectonically active convergence zone of the Nazca and the South American plates, is characterized by uplifted terraces.

THE CHANGING SEA LEVEL

The scarp on Montague Island along the Hanning Bay Fault was formed by the Alaskan earthquake in 1964. Dead mussels and barnacles cover the rocks on the shore of this island. Uplifted by the earthquake, land that was once below the high tide level now supports grass.

tremors ever recorded in North America. Up and down the Alaskan coast, blocks of land up to kilometers across rose or subsided as much as 2 m. At one fishing port, bedrock rose more than 6 m above sea level, trapping fishing boats on land and leaving the dock pilings high and dry. Offshore, nearby islands uplifted as much as 12 m; and on the adjacent continental shelf, a section of the seafloor dropped more than 15 m—all this happened in a matter of minutes.

A similar sudden shift in sea level occurred in Napier, a port located on the east coast of the North Island of New Zealand, which straddles a major convergence zone. In the morning of February 3, 1931, a strong earthquake nearly destroyed this community. The port's shallow bay was uplifted more than 2 m, well above the high tide level. Fish flopped about and boats were stranded. Today, what was formerly the bay is farmland and the site of the municipal airport. An artificial harbor now serves the community, and a narrow channel draining the adjacent upland is all that remains of the uplifted bay.

Severe earthquakes that permanently change local sea level are fairly common wherever plates collide. They can be expected to occur along the convergent margin of the west coasts of North and South America, along the eastern Mediterranean through to the Middle East, and up and down the western edge of the Pacific Ocean plate.

CLIMATIC FLUCTUATIONS

Seasonal and other cyclical fluctuations in climate also affect regional sea levels. Mean sea levels along coasts throughout the world typically show seasonal differences of 10 to 30 cm (centimeters). Spring is generally the season of lowest sea levels and fall the season with highest sea levels. The pattern is the same in both the Northern and Southern hemispheres, following the alternating pattern of the seasons. Closer to the equator, where the seasonal shifts are more subtle, mean sea level is more constant.

The simplest and most easily predictable of the seasonal sea level fluctuations are a direct result of annual weather patterns. As the Sun moves overhead across the latitudinal belts of the Earth, surface water temperatures rise or fall; wind patterns and wind intensities change as a consequence. Regional mean sea levels experience a modest change in response to these seasonal wind differences, because, as water is pushed by the wind, it piles up in the direction of the wind and leaves a small depression behind.

Although the overall pattern is a predictable one, it is interesting to note that the seasonal differences vary greatly from one area to another. In the Bay of Bengal, for example, mean sea level varies by as much as 100 cm per year, because of the seasonal extremes of the dry winters and the wet monsoon-drenched summers combined with the reversals in wind patterns. During the summer, the winds from the south hold the tremendous runoff from the north in this embayment. In other bodies of water, local wave and current patterns produce variations from the expected seasonal patterns.

Shifting wind patterns are not the only climatic feature affecting regional sea levels. Almost everyone knows that the volume of a given mass of water changes as the water's temperature changes. But it is not so well known that the seasonal changes in seawater temperature produce volume changes large enough to affect regional sea levels. In the Chesapeake Bay, on the United States east coast, the sea level at Annapolis, Maryland, is highest in the summertime because the warm Gulf Stream flows close on shore during that season and warms the local waters. Across the continent, at Neah Bay, Washington, the sea level in the summertime is at its lowest because the cold California Current flows down the coast from the north at that time. Moreover, as the California Current travels from high in Canada

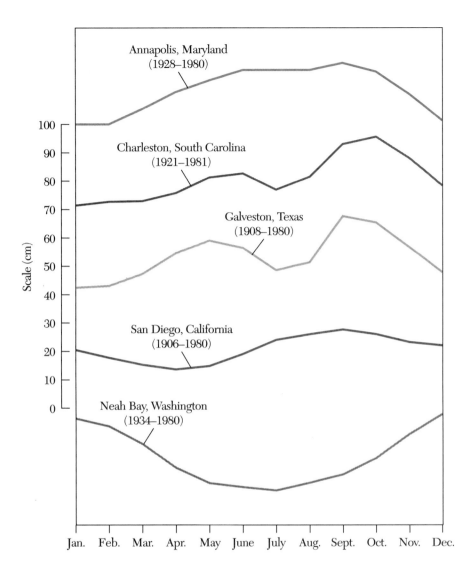

Annapolis, Maryland
(1928–1980)

Charleston, South Carolina
(1921–1981)

Galveston, Texas
(1908–1980)

San Diego, California
(1906–1980)

Neah Bay, Washington
(1934–1980)

The individual patterns of the annual cycles of sea-level changes in various locations become evident when the average monthly tide gauge records are graphed. The differences between sites result from a combination of variations in coastal configuration, water temperature, weather patterns, and major current systems.

to mid-California, some of the water in the current is always veering west, out to the Pacific Ocean, driven by winds deflected to the west by the Earth's rotation. Colder subsurface waters upwell to replace the lost water, contributing further to the region's unusually low sea level. Neah Bay's highest mean sea level occurs in winter, not because the waters are warmer, but because frigid Arctic winds generate coastal storms that pile up water in the bay.

E. C. LaFond of the Scripps Institution of Oceanography monitored the sea level cycle of the southern California coast for several decades and found a strong correlation between the seasonal fluctuations in sea level and the surface temperature of the water. He also found a striking correspondence between the sea level cycle and the pattern of erosion along the coast. High sea level and high energy conditions during the winter months were, as expected, coincident with the greatest rates of coastal erosion—a good example of how some commonly overlooked cyclic changes in sea level can significantly influence coastal erosion.

An anomalous sea level cycle occurs every four to seven years over virtually the entire South Pacific Ocean in response to the appearance of El Niño, a warm current flowing off the western coast of South America. Fishermen have named the current "boy-child"—a name referring to Christ—because the phenomenon of El Niño tends to appear in late December. Normally at this time—that is, summertime in the Southern Hemisphere—the cold Peru Current flows northward up the coast. It is accompanied by a wind-driven upwelling of even colder nutrient-rich waters from below at the point where the Peru Current veers west into the

The coincidence between monthly average sea level and water temperature in Southern California is apparent, as is their relationship to the dominant oceanic conditions throughout the year.

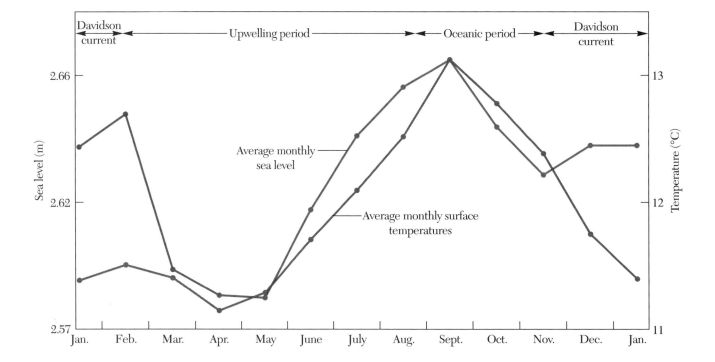

Pacific. This upwelling of fertile waters supports one of the world's most important fishing grounds. (Incidentally, the path of the Peru Current is almost a mirror image of the track of the summertime California Current.)

At irregular intervals of four to seven years, the west-to-east trade winds diminish, allowing a warm current to move south and drive the Peru

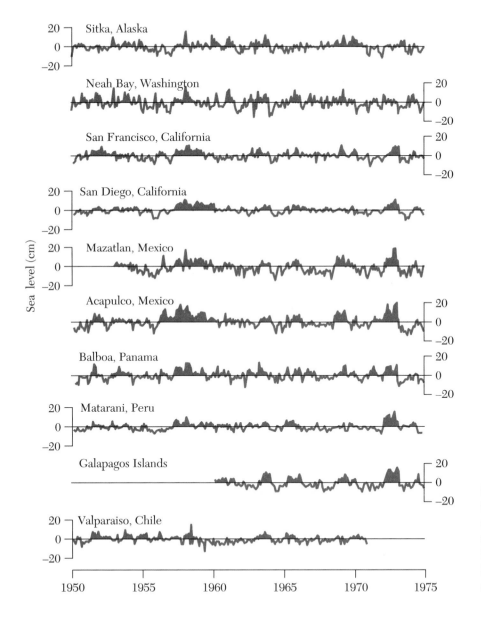

Sea-level variations recorded throughout the Pacific Ocean over a 25-year period show anomalies that have been attributed to an El Niño effect. Similar patterns can be seen in all of the eastern Pacific locations charted.

Current into the Pacific prematurely. This warmer current is the El Niño. When it strays into South American coastal waters, its thick surface layer of warm water acts as a barrier to the upwelling of the cold, nutrient-laden waters below. It may stay in this southern location for a year or more, causing fish populations either to migrate to other feeding grounds or to die. Wintertime temperature differences at the sea-air interface bring on torrential storms, which cause flooding and coastal erosion, and changes in wind and current patterns. Whenever El Niño flows south, the sea level rises all along the western South American coast because of El Niño's climatic effects and because its warm waters have a greater volume than the colder Peru Current.

While El Niño is active, the changes in wind and current patterns produce very long waves that move across the entire Pacific Ocean. As in all onshore wind conditions, trade winds normally cause a pileup of water along coastal areas of the western Pacific. The diminution in the trade winds that is associated with El Niño releases the piled-up water in the form of a very long wave moving on the ocean surface.

Paul Komar of the School of Oceanography at Oregon State University studied one of these waves, which developed during the 1982–83 El Niño. He traced its movement by collecting and collating sea level data from the tide gauge records of several coastal stations around the Pacific Ocean. His calculations showed that the wave traveled east at a speed of about 75 km per day and maintained a height of 30 to 40 cm from its point of origin to its point of arrival at the western North American coast. There was extreme erosion along the California and Oregon coasts that winter. Severe storms caused some of the damage, but some must be attributed to the high sea level brought about by the El Niño-induced wave.

Regional Subsidence due to Compaction and Fluid Withdrawal

In parts of the coastal zone in east Texas and the Mississippi River delta in Louisiana, sea level is rising 9 to 10 mm per year because the land itself is sinking. This rate of sea level rise is the highest in the United States and is approximately three times the global average. About 6 to 7 mm of this annual rise is due to subsidence caused by a combination of compaction of delta sediments and the withdrawal of fluids from the coastal zone.

In the lower Mississippi River delta, huge quantities of fine sediment— as much as 1.6 million metric tons per day—arrive at the river's mouth. These sediments trap water as they settle, and the resulting mud, at places up to 90 percent water by volume, accumulates in the active delta region. The weight of all this mud compresses the underlying sediments and eventually drives out its own water. As the water leaves, the land surface subsides. Because the remaining sediment from the mud adds less volume than was lost from compaction of the underlying sediments, the net effect is a regional rise in sea level. The subsidence of the Mississippi River delta has been accelerated during the past few decades by the many dams and other controls placed by humans on the river, which have slowed the influx of new sediment while older sediment has compacted.

Compaction can take place in any large and rapidly growing delta. The Amazon River delta is another river mouth receiving great quantities of sediment, and the rate is increasing as the rainforests disappear and erosion accelerates. Some subsidence is also present at the mouth of the Amazon, but its remoteness limits the availability of information on its rate.

Compaction is not only responsible for subsidence in delta regions throughout the world; it also affects peat bogs, marshes, and other organic-rich sediment accumulations that contain large volumes of water.

Whereas the natural process of compaction has been occurring for many millions of years, a significant part of the northern coast of the Gulf of Mexico has been subsiding because of the withdrawal of large volumes of fluid by various human activities. Nearly 100,000 wells have extracted huge

The marshes in this portion of the Mississippi Delta are drowning as the sea level rises at a rate of nearly 10 mm per year.

volumes of oil and natural gas from the Mississippi River delta and the nearby coastal zone of the Gulf of Mexico. Large quantities of water also have been extracted for domestic and industrial uses. One result of these activities is that some of the land near Galveston, Texas, has sunk nearly 2 m in this century. To lessen the threat of further subsidence, water is now being pumped back into the ground.

Another human impact on the coastal zone is not as easily remedied. The practice of building high-rise city centers on unstable sediments has provoked settling in portions of the inland port cities of Houston and New Orleans, which are now below sea level. Dikes and levees have been constructed to keep seawater out.

If sea level in the lower Mississippi delta continues to rise at the rate of 9 to 10 mm per year and if this rate is sustained for 50 years, the sea level could rise by as much as 0.5 m, enough to submerge large areas of coastal wetlands. This total sea level rise is the combination of 3 mm from global rise, another 1 to 2 mm from fluid withdrawal, and the remaining 5 to 6 mm from compaction.

ISOSTASY: SUBSIDENCE AND REBOUND OF THE LITHOSPHERE

The sinking Mississippi River delta illustrates the phenomenon of regional subsidence, but broader regions of the continental lithosphere can also sink under a heavy load. Geological evidence indicates that, over several thou-

A cross-section from Trondheim, Norway, to Helsinki, Finland, shows the subsidence and the rebound associated with glacial development and subsequent melting of the Fennoscandian ice sheet. The area in which the land remains below sea level is the Baltic basin.

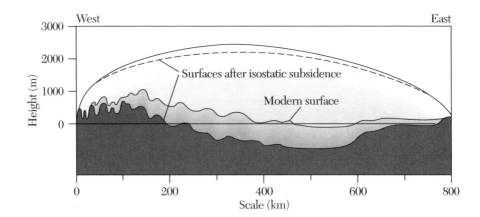

sand years, the weight of a growing ice sheet can warp the continental lithosphere downward by as much as 200 to 300 m, depending on the thickness and density of both the ice mass and the underlying lithosphere. When the glacier melts, the lithosphere rises again, or rebounds, as the weight of the ice is removed by runoff from the landmass. These shifts in the weight of the continental plates have taken place numerous times as glaciers advanced and retreated.

Adjustments in the relative position of the lithospheric crust in the asthenosphere are called isostatic adjustments. Isostasy is the condition of equilibrium of the Earth's crust that is achieved by a constant rebalancing of the forces that tend to elevate the lithosphere against those that tend to depress it.

The loading and unloading—and subsequent sinking and rebounding—of the lithosphere is only one factor affecting isostasy. A change in the density of the lithosphere also will cause isostatic crustal movement and will produce corresponding changes in sea level. Young, hot lithospheric material that is produced at the rift zone of the oceanic ridge system is relatively low in density. Over a period of several millions of years, this lithosphere cools, contracts, becomes denser, and slowly subsides into the asthenosphere. Because most of this activity takes place in the ocean basin, the rise in sea level above it is imperceptible.

At the few regions on Earth where coastlines are close to diverging plates, the subsidence of the lithosphere can produce a rise in sea level and a slow inundation of the land. The coasts of both the Red Sea and the narrow Gulf of California lie near the boundary between diverging tectonic plates and are candidates for this type of sea level rise.

This bluff along the coast of Peru contains Quaternary sediments that were deposited below sea level but are now about 100 m above sea level as the result of tectonic activity. A close-up (*right*) reveals the variety of sediments deposited.

Isostatic adjustment of the Earth's crust is also produced by the prolonged accumulation of sediments and rocks on a portion of the lithosphere. This process takes place in sedimentary basins or along thick prograding coastal plains such as the Gulf and Atlantic coasts of the United States. In both cases, a thousand or more meters of sediment have accumulated over tens of millions of years and have caused the entire plate to subside—carrying the sediments down with it.

The same phenomenon is associated with the thick accumulations of volcanic rocks, especially on islands in the Pacific Ocean. The Aleutian islands also show some subsidence produced by loading, but the amount of sea level change due to tectonic activity associated with a subduction zone is so large here along an active plate boundary that it masks that due to subsidence caused by isostatic adjustment.

CHANGES IN THE VOLUME OF THE WORLD OCEAN

A global—or eustatic—change in sea level can come about in three ways: by adding or reducing the amount of water in the entire world ocean; by changing the volume of the ocean basins; or, more subtly, by changing the mean temperature of the world ocean so that its volume increases or decreases as the seawater expands or contracts. In any case, when the volume of water changes, it causes eustatic change in sea level.

The volume of the ocean basin changes when tectonic forces drive continents apart or pull them closer, widening or narrowing the world oceans. These changes are extremely slow, requiring tens of millions of years rather than the tens of thousands required by the other two processes. Because their impact is far more immediate, the rest of the chapter considers changes in sea level due to changes in the amount of water or in the expansion of water.

A worldwide and long-term change in climate can bring about both of these conditions simultaneously: A cold climate forms ice sheets and thereby traps water in the glaciers while at the same time lowering the temperature of the ocean. Both conditions reduce the volume of the world ocean. Conversely, a warm climate melts the glaciers and returns the water to the ocean and at the same time raises the ocean's temperature. These conditions increase the ocean's volume and, therefore, raise global sea level.

Drastic temperature changes are not required to produce drastic changes in sea level. A rise or fall in the mean annual temperature of only 1–2°C has a profound effect on both the volume of ice retained on the surface of the Earth and the volume of the water in the world ocean. During the last ice age, thick ice sheets covered about 30 percent of the land area of the Northern Hemisphere. The ocean receded, its waters bound up in ice or contracted with the cold, and nearly all of the continental shelf was exposed. These conditions were produced with a mean annual temperature only 2–3°C lower than it is today. If the current trend toward a warmer global climate continues and the mean annual temperature increases by only a few degrees, the entire process will reverse—the ice sheets will melt and the ocean will encroach upon the continents until most of the port cities are under water.

Insufficient data have made it difficult, however, to access the pace and direction of global climate and sea level changes over the short term. Accurate records of sea level have been kept for little more than a century, and our weather records in most parts of the world do not reliably extend back even that far. The great sea level changes of the past, as recorded in the advances and retreats of the glaciers, occurred in cycles of tens of thousands of years or more. Our hundred-year-old records, therefore, cannot be used to make valid predictions.

Nevertheless, the recent rise in global temperature has forced us to take note of a possible human-generated absolute rise in sea level. The still-accelerating release into the atmosphere of carbon dioxide and other

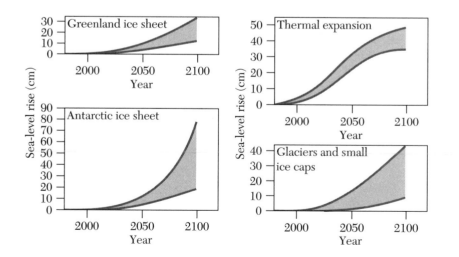

A recent study predicted the contributions to sea-level rise that would be made by various factors during the next century. The projections indicate melting of the ice caps and glaciers is the dominant factor, with thermal expansion having the least impact over the period.

In projecting a 3°C increase in global temperature by the year 2030, a United Nations Environment Program study estimated the relative amount of warming that would be caused by various gases released into the atmosphere. Increased carbon dioxide is almost as much a source of warming as all other factors combined. (CFC refers to chlorofluorohydrocarbons.)

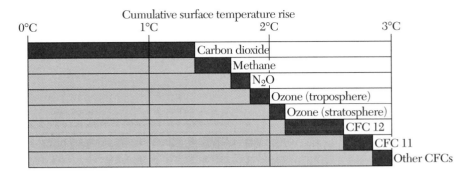

Cumulative surface temperature rise

greenhouse gases has some climatologists projecting a global warming of 3°C by the year 2030. Such a rise in mean annual temperature could melt enough of the ice cover in Greenland and Antarctica to raise the global sea level by 30 to 40 m in a few centuries. Even here, however, we may be trying to superpose a short-term database on the vast periods of cyclicity. The current increase in carbon dioxide might just be part of a larger cycle that predates civilization and will decline by itself. But it is hard to discount the obvious contribution to global warming being made by our current high levels of combustion and our destruction of photosynthesizing plants. It would be prudent for us to take careful note of the warming and melting patterns of the past and their dramatic effects on global sea level.

ADVANCE AND RETREAT OF ICE SHEETS

Geologic evidence indicates that during the Quaternary Period, which began about 2 million years ago and continues today, eustatic sea level changed more and faster than ever before in the Earth's history. The driving force behind these sea level changes has been the growth and shrinkage of continental ice sheets and the polar icecaps in Antarctica and Greenland. Until 20,000 years ago, when the Pleistocene Epoch of the Quaternary came to a close, thick ice sheets repeatedly covered and withdrew from most of Europe, northern Asia, and North America down to the Missouri and Ohio river valleys. Except for the large ice sheet over Antarctica, their incursions in the Southern Hemisphere were limited because of the paucity of land in the high latitudes.

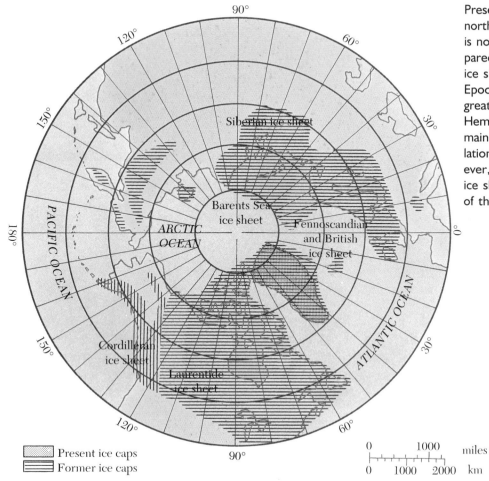

Present-day coverage of the northern part of the earth by ice is not very significant when compared to the distribution of glacial ice sheets during the Pleistocene Epoch when they reached their greatest extent. In the Northern Hemisphere, only Greenland remains a major site of ice accumulation; it is extremely small, however, compared to the Antarctic ice sheet on the southern portion of the globe.

Early in this century, the record of the Pleistocene Epoch that is preserved on the continental landmasses was interpreted to show four ice ages during which glaciation alternated with melting. Each of these glacial and interglacial periods lasted at least tens of thousands of years. In North America these periods of glacial advance are named, from oldest to youngest, the Nebraskan, Kansan, Illinoisan, and Wisconsinan, after the states where the respective ice sheets deposited considerable sediment. Europe and other parts of the world have their own terminology. There is land-

Chronology of Quaternary Glaciation

Epoch	Glacial Stage	Interglacial Stage
Holocene		
	Wisconsinan	
		Sangamonian
	Illinoisan	
Pleistocene		Yarmouthian
	Kansan	
		Aftonian
	Nebraskan	
Pliocene		

based evidence that the glacial stages culminated 630,000 before the present (B.P.) (Nebraskan), 430,000 B.P. (Kansan), 150,000 B.P. (Illinoisan), and 20,000 B.P. (Wisconsinan).

For the first half of the twentieth century, this interpretation was accepted and taught, not only in North America but throughout the rest of the world as well. However, as more and more information about the ocean floor was obtained through sediment cores, it became apparent that many more Pleistocene glacial events have taken place. They were not noted earlier because their traces have been masked on land by the deposits of the larger glacial incursions.

This discovery is primarily the result of new techniques for investigating ancient climatic conditions on the Earth. Foremost among these is the use of oxygen isotopes, that is, variants of the oxygen atom. Oxygen typically has an atomic mass of 16, but there is also a heavier isotope, oxygen-18. Both isotopes behave the same chemically and both are incorporated in the skeletons of organisms as part of the compound calcium carbonate ($CaCO_3$). In 1947, Harold C. Urey of the University of Chicago theorized that the relative concentrations of the two oxygen isotopes taken into the skeleton of organisms were dependent on the temperature of the ocean water in which they lived. He then reasoned that this water temperature could be discovered by using a mass spectrometer to determine the $^{16}O/^{18}O$ content of the calcium carbonate in the organism's skeleton.

In 1955 Cesare Emiliani, a young associate of Urey, analyzed the skeletons of planktonic foraminifera from several deep-sea cores. These floating, single-celled protozoa are very abundant in deep-sea sediments. From these cores Emiliani determined that there were numerous significant temperature fluctuations in ocean waters over a period of only 300,000 years, an interval far shorter than the total time of the Pleistocene glaciation. The fluctuations were assumed to be in response to climatic changes associated with glacial activity and were in marked contrast to the Pleistocene glacial history derived from land-based studies.

Furthermore, the cycles shown by the isotopic data from the fossils in Emiliani's sediment cores corresponded to a periodicity that had been predicted many years earlier by Mulitin Milankovitch, a Yugoslavian astronomer. Milankovitch's theory of climatic changes was based on cycles in the radiation received by the Earth as it tilts relative to the Sun. For the period prior to about 700,000 B.P., Milankovitch's cycles had a periodicity of 41,000 years; from 700,000 B.P. to the present, the periodicity was 100,000 years. Therefore, Emiliani's fluctuations were assumed to be a response to changes of similar periodicities in major oceanic circulation patterns, which were in turn a result of the radiation cycles noticed by Milankovitch.

The present North American coasts are the products of only the most recent of these glacial cycles that created the conditions molding the world's shorelines. The last great advance of glaciers, the Wisconsinan ice age, began about 120,000 years ago and persisted for over 100,000 years. Since then we have been in the Holocene (formerly called the Recent) Epoch, a period characterized by glacial melting. The Antarctic and Greenland ice sheets of today are remnants of that Wisconsinan advance. This most recent period of overall melting and related warming has been interrupted by a few "little ice ages." These shorter episodes are really only extended periods of slightly colder weather, some of which have occurred during recorded history. The most prominent was chronicled in Europe between about 1450 and 1850.

If we calculate the difference in volume between the Wisconsinan ice sheets and those remaining today, we can determine how much ice melted and how this volume of released water affected global sea level. We can then extrapolate forward in time and estimate future sea levels and their consequences, assuming the present rate of global warming and continued melting of all remaining ice sheets.

First, we must estimate the surface area and depth of the previously existing ice sheets. The surface area covered by the Antarctic ice sheet during the Wisconsinan advance was nearly 14 million km^2; today it is

Area of Present and Maximum Extent of Ice Sheets

Region	Present Area (km^2)	Maximum Area (km^2)
Antarctica	12,535,000	13,800,000
Greenland	1,726,400	2,295,300
Laurentide Complex	147,200	13,386,900
Scandinavia/United Kingdom	3,800	6,666,700
Rocky Mountain area	76,800	2,610,100
Asia	115,000	3,951,000
European Alps	3,600	37,000
South America	26,500	870,000
Australasia	1,000	30,000
Total	14,635,300	43,647,000

Modified after Andrews, J. T., 1975; data from Flint, R. F., 1971.

about 12.5 million—not really much smaller. The North American Laurentide ice sheet east of the Canadian Rockies is lower in latitude and therefore subject to more melting than the Antarctic icecap. It once covered more than 13 million km^2 of land and now covers only 147,200. The total surface area covered by ice 20,000 years ago was more than 44 million km^2, and it is now just under 15 million, one-third of its former extent.

It is more difficult to determine the volume of the Wisconsinan ice sheet than it is to determine its surface area. Perhaps the best way to approach the problem is to use the Greenland ice sheet as a model, because the Antarctic ice, although more than 2 km thick in some places, rests on rugged topography, which causes the ice sheet's thickness to vary greatly. Moreover, much of the Antarctic topography is unexplored. The dimensions of the Greenland ice sheet, however, are well known because of extensive petroleum exploration surveys and many scientific and military studies. Glaciologists have estimated that the Greenland ice sheet holds about 2.5 million km^3 of ice.

By estimating surface areas and average thicknesses for various latitudes, we conclude that 75 million km^3 of ice were contained in the vast glaciers of the Wisconsinan ice age. Therefore, about 50 million km^3 of ice have melted since then—a volume roughly equivalent to 20 times the amount of water locked in today's Greenland icecap. Allowing for a 10 percent decrease in volume when ice turns to water, we estimate that 45 million km^3 of water were returned to the world ocean.

The mammoth shift of water mass from the continents to the ocean had—and is still having—tremendous repercussions. As the mass of the ice sheets was gradually removed from the backs of the continents through melting, the continental lithosphere beneath the departing load began to rise, or rebound, in an isostatic adjustment. If, as is likely, some regions were covered with ice 3000 m thick at the peak of the ice age, the rebound when the ice melted would have been about 1000 m. [This figure is based on the assumption that 3000 vertical meters of ice at a density of 0.9 grams per cubic milliliter (about one-third the density of the continental lithosphere) is equivalent to about 1000 vertical meters of lithosphere.] The pace of the rebound has been slow and uneven; and in some places, it is still going on.

Meanwhile, in the world ocean, the added mass pouring in from the continents caused a sinking of the ocean floor, which has a much younger and thinner lithosphere than the continents have. Nevertheless, the tremendous increase in the volume of water raised sea level despite the subsidence of the ocean floor. Taking into account the uplift of the continents, the subsidence of the ocean floor, and the increase in the volume of ocean water, we arrive at an estimate of 120 m for the long-term net rise in global sea level, with local variations due to differences in continental rebound.

CONTINENTAL REBOUND

The continental rebound was slow as the Wisconsinan glacial ice melted gradually, and the plastic asthenosphere responded slowly. Nevertheless, most of the isostatic adjustment of the underlying crust was accomplished before the ice sheet was completely gone, because the weight of the remaining ice was no longer sufficient to cause significant subsidence. In some places, however, where the ice was thickest, the rebound continues. Significant uplift of continental mass is taking place today in Scandinavia, in the Hudson Bay area of Canada, and perhaps in Argentina and southern Chile.

Along the Scandinavian coast, the mean annual sea level drops about 9 mm per year, an observation indicating an uplift of continental lithosphere. When we take into account the annual global sea level rise of 2 to 3 mm, the net rebound is 11 to 12 mm per year. On the basis of this figure, we calculate that the continent has rebounded 55 to 60 m since glaciers

Mansel Island in northeastern Hudson Bay was submerged during deglaciation. As the land isostatically rebounds, glacial deposits are reworked into a linear series of raised beaches, as shown by the northwest-southeast trending light colored zones on the photograph.

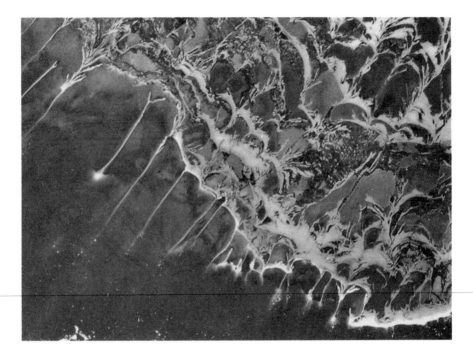

receded from Scandinavia 5000 years ago. The total rebound in the Hudson Bay area of Canada is even more. In aerial photographs of the northern Scandinavian coast, and of other coasts of active uplift, one can see a parallel series of ridges marking the old shorelines and abandoned beaches as the relative sea level dropped and the land emerged from the ocean.

The rate of active isostatic uplift diminishes in a southward direction in the Northern Hemisphere, because the ice was thinner and melted more rapidly in the southern latitudes. This differential is seen in the stratigraphic record of the New England coast: past shorelines and the features preserved in them have tilted upward to the north. The mean annual sea level also reflects the uplift; it drops several millimeters a year in Maine but holds steady in New York. The Great Lakes, which owe their origin to glaciers, show even clearer signs of a north-to-south tilt; the ancient shorelines on the northern sides of all the lakes are several meters higher than the same-age shorelines on their south sides. Lake Michigan is gradually getting shallower at the north end and deeper in the south. If the present rate of rebound persists, then in about 3700 years Lake Michigan will empty into the Mississippi River system rather than past Detroit through the other lakes into the St. Lawrence Seaway, as it does now.

THE CHANGING SEA LEVEL

The Holocene Rise in Sea Level

The Holocene Epoch is characterized by the melting of the last Pleistocene ice sheet, the Wisconsinan, and the subsequent rise in global sea level. We would most like to determine the following facts: When did the rise begin? How fast has it proceeded? How long will it last? How high will it go before it reverses? In other words, we wish to determine the sea level curve of the past and extrapolate the next segment of the curve. Unfortunately, geologists, oceanographers, and climatologists who study sea level fluctuations cannot agree among themselves about what the past projection should look like—an understandable situation, given the complexity of the problem.

It is generally agreed that sea level was at its lowest position about 18,000 to 20,000 years ago, when the ice sheets of the Wisconsinan ice age reached their maximum development. The lowest level, called the lowstand, is deduced by uncovering evidence of the oldest drowned shoreline on the continental shelf. This evidence might take the form of beach sand, marsh deposits, drowned wave-eroded platforms, or almost any indication of an old shoreline, such as fossil plant or animal material.

As noted earlier, other factors such as tectonic uplift and subsidence must also be taken into account before deciding the vertical position of the lowstand. Most researchers have placed the lowstand depth about 120 m below present sea level and have agreed that sea level rose rapidly until about 6000 to 7000 B.P. The mean annual rise during this period was

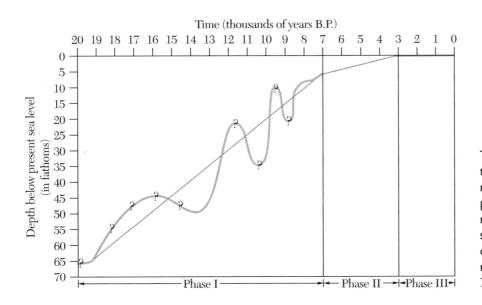

This simplified sea-level curve for the Holocene period shows the rate of rise divided into three phases: rapid rise, slow rise, and relative stability. The blue curve shows fluctuations that possibly occurred during the period of rapid rise that lasted until about 7000 years B.P.

nearly 10 mm, fast enough for the sea to cover most coastal cities in a thousand years.

During the period of rapidly rising sea level, the shoreline moved so rapidly that the sandbars and barrier islands that protect so many of our coasts today had no time to develop. The lack of a stable shoreline coupled with moderate to high tidal ranges produced tide-dominated coasts with widespread estuaries and tidal flats. In areas such as Australia and New Zealand, sea level reached its present position about 6500 to 7000 years ago, at the end of the period of rapid rise. In most parts of the world, however, sea level continued to rise, but at a much reduced rate, because glacial melting slowed. During the period of slower rise, shorelines became more stable and waves became the dominant factor for coastal development. Under these conditions, beaches and barrier islands were formed.

There is disagreement about the positions of sea level in much of North America during the past 3000 years. Some researchers believe it has been stable at the present position; some believe it has moved above and below its present position; and many think sea level has gradually risen during the period, but only by about 3 m.

CURRENT AND FUTURE SEA LEVEL CHANGES

The depletion of the ozone layer in the Earth's atmosphere and apparent global warming have had noticeable effects on the sea level of the global ocean. Approximately 100 years' worth of reliable data from tide gauges around the world indicate that eustatic sea level is rising at an increasing rate. Good records come from tide gauges that are mounted on a stable surface, preferably bedrock. Pier pilings, for example, are not very reliable because they can settle or they may shift position due to the scouring of tides or waves.

Much of our knowledge of recent trends in sea level is due to the diligent and long-time efforts of Steacy Hicks of the National Ocean Survey. Hicks and his colleagues have studied and interpreted data from hundreds of tide stations.

The data from both coasts are influenced by their tectonic settings. The tectonically more stable east coast has experienced a rise in sea level that ranges from a relatively slow rise in the north, where isostatic rebound is still going on, to a more rapid rise to the south, where this phenomenon is absent. The short-term ups and downs are considerable, and in New England there is a greater than average range in the average rate of rise. The west coast, in part a plate convergence area, has a greater range in sea level

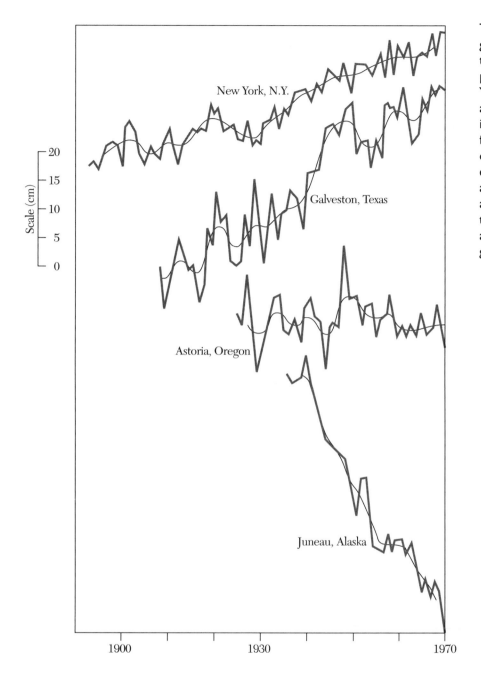

Tide gauge records have been gathered at various locations along the United States coast over the past several decades. The New York record is close to the global average; the Galveston rate of rise is high due to fluid withdrawal from and consequent subsidence of the land; Astoria, Oregon, is essentially stable; and in Anchorage, Alaska, the sea level is actually falling. The latter two locations are influenced by tectonic activity at a zone of plate convergence.

A map of the continental United States compares mean annual rates of sea-level rise as determined by tide gauge data. The tectonically stable Atlantic coast has modest rates of rise; the Mississippi Delta has a very high rate of rise; and the Pacific coast shows both rising and falling sites because of its location on a converging plate boundary.

changes. Low rates of sea level rise are found along the lower west coast, but in Alaska the situation is generally reversed. Many locations in Alaska actually show a relative drop in sea level, because of rapid tectonic uplift of the continental edge—up to 14 mm per year.

In the last century all the coasts of the global ocean have experienced a net rise in sea level, except for one huge area, the coasts around the Pacific Ocean. These coasts are mostly convergent coasts where sea level changes caused by tectonic conditions mask eustatic changes in sea level.

It should be noted, however, that one century of data for sea level changes is insufficient to accurately predict long-term sea level trends. During the past several thousand years, perturbations in sea level have occurred as climatic conditions varied within the long-term warming and melting cycles. These small deviations were sometimes a century or more in length. A good example of such a deviation took place a few hundred

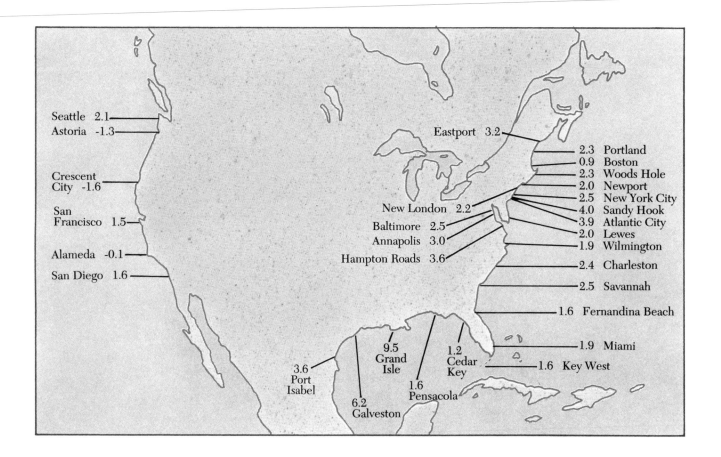

Seattle 2.1
Astoria -1.3
Crescent City -1.6
San Francisco 1.5
Alameda -0.1
San Diego 1.6

Eastport 3.2
2.3 Portland
0.9 Boston
2.3 Woods Hole
2.0 Newport
2.5 New York City
4.0 Sandy Hook
3.9 Atlantic City
2.0 Lewes
1.9 Wilmington
2.4 Charleston
2.5 Savannah
1.6 Fernandina Beach
1.9 Miami
1.6 Key West

New London 2.2
Baltimore 2.5
Annapolis 3.0
Hampton Roads 3.6

3.6 Port Isabel
6.2 Galveston
9.5 Grand Isle
1.6 Pensacola
1.2 Cedar Key

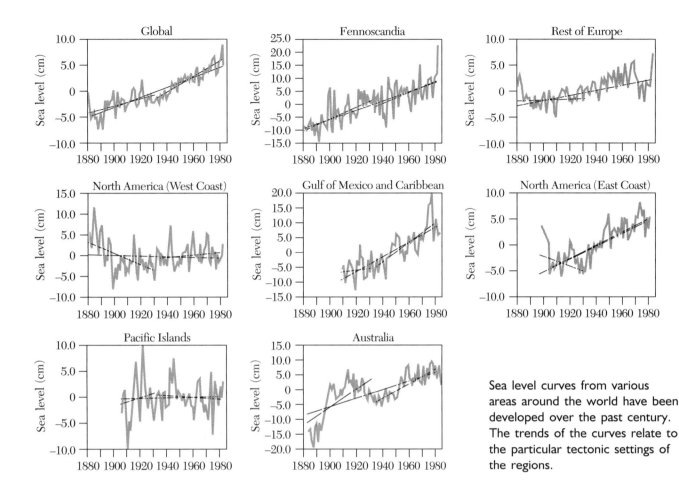

Sea level curves from various areas around the world have been developed over the past century. The trends of the curves relate to the particular tectonic settings of the regions.

years ago in what is known as the "little ice age" (1450 to 1850). During that time, there was a marked change in climate; the existing ice sheets appeared to increase in size, and the global sea level fell slightly. This reversal of the long-term trend was best documented in Europe, where an extensive population was present to record the events associated with it. Similar but smaller scale perturbations have been recorded in recent times by carefully observing individual glaciers and by correlating the observations with weather records. It has been shown that glaciers advance and retreat in accordance with weather records and that the periodicity ranges from years to decades. Factors that contribute to these perturbations include variations in sunspots, shifting of oceanic currents, and slight changes in Earth–Sun positions.

IMPLICATIONS FOR COASTAL ENVIRONMENTS

The coast as we know it today represents one of the youngest elements of the Earth's crust. Most of the coast we see now goes back only a few thousand years; the total history of the Earth covers about 5 billion years. The coast represents one of the fastest changing parts of the Earth's surface.

Changes in sea level either directly or indirectly affect virtually all of the sediment accumulations and landforms in the coastal zone. As the shoreline moves, it either exposes or inundates the coastal areas and, in so doing, causes the character of the coast to change. Additionally, the position of the shoreline strongly influences coastal processes, such as waves, tides, and currents, that act to shape the coastal environments.

During the period of significant human occupation on Earth, eustatic sea level has changed by less than 3 m. As a result, we are often unaware of the fragile nature of the coast, and we have come to take its relative stability for granted. We must recognize, however, that the nature of the sea is dynamic, not static. A continued rise in sea level of 3 to 4 mm per year for a century or more will devastate many densely populated parts of the world. We need only look at the Mississippi River delta region, where an

Ten Countries Most Vulnerable to Sea Level Rise

Country	Population ($\times 10^6$)	Per Capita Income (in dollars)
Bangladesh	114.7	160
Egypt	54.8	710
Gambia	0.8	220
Indonesia	184.6	450
Maldives	0.2	300
Mozambique	15.2	150
Pakistan	110.4	350
Senegal	5.2	150
Surinam	0.4	2360
Thailand	55.6	840

From Ince, M., 1990.

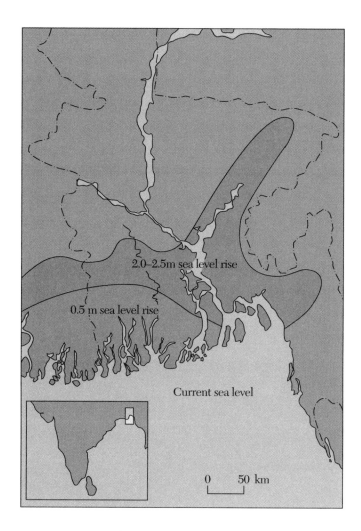

The Ganges–Brahmaputra Delta area on the Bay of Bengal is extremely susceptible to flooding if the sea level rises as predicted. The map shows the extent of projected flooding that is likely to occur if the sea level rises 0.5 m and 2.0–2.5 m.

annual rise of 9 mm now causes a loss of over 40 acres of coastal land each month, to see the damage that can result from the rising global ocean.

A more drastic situation is unfolding in Pakistan, a poor country inhabited by 115 million people. Situated in the delta area of the Ganges-Brahmaputra rivers, it is flooded nearly every monsoon season. Many lives are lost, and even more individuals are displaced from their homes each time this happens. If this area experiences as much as a 0.5 m rise in sea level over the next 50 years, more than 1000 km^2, or 0.1 percent of the country's area, will be inundated.

3

Processes That Shape the Coast

Thus far we have considered the large-scale and long-term forces that shape thousands of kilometers of coast. We now turn to short-term, day-to-day agents, which can be quite localized in their effects, but which give each kilometer of the coast its own special character.

There are three quite different types of processes that influence the configuration of the coast: physical, chemical, and biological. Foremost are the physical processes—the waves, the tides, and the currents that the waves and tides generate. The sheer force of these three processes wears down the coast in some places and builds it up in others, transports sand and shell, molds barriers and spits, shapes coastal bays, and makes rivers change their courses.

In subsequent chapters we will explore the effects of these three sculptors on regional and local coasts—the second-order and third-order features of coastal variation in the Inman and Nordstrom classification scheme. In this chapter we will examine the physical processes themselves.

Breaking waves release enormous energy that continuously transforms the coast.

WAVES AT THE COAST

All waves are disturbances of a fluid medium through which energy is moved. Sound and light, for example, travel as waves through space which is perturbed but not displaced. Our concern here is with the wave that travels on the ocean surface, on the interface between the ocean and the atmosphere. When this interface is disturbed, a wave is produced. As we will see, the water within the waveform is perturbed but does not advance across the surface with the wave.

The three primary forces that produce surface waves are wind, earthquakes, and the gravitational attractions within the Sun, Moon, and Earth system. The wind is the most familiar and easily observable of these phenomena, and it generates nearly all the waves we see.

To facilitate our discussion, we will review the standard terms for the different parts of a wave. The crest is the high point of a wave, and the trough is its low point. The wave height is the vertical distance from trough to crest. The length of the wave, called the wavelength, is the horizontal distance from crest to crest or from any point on the wave to the same point on the next wave. The steepness of a wave is the ratio of its height to its length (H/L). When the steepness reaches a ratio of 1:7, the crest of the wave topples over. Thus, a wave that is 7 m long can be no higher than 1 m; if it becomes any higher, it is unstable and breaks.

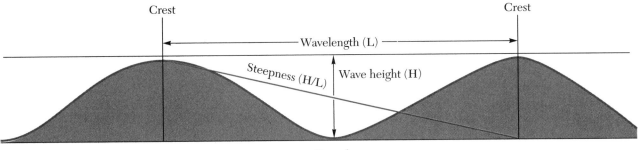

Each wave that develops on the surface of the water has a crest and a trough. The relationships between the crest and trough of a wave—the wavelength, wave height, and steepness—differ depending on the conditions.

Waves are formed by a disturbing force; for most waves, this force is produced by the friction of the wind moving across the water. The force might also be an earthquake for a seismic sea wave (tsunami) or an abrupt change in atmospheric pressure for the back-and-forth sloshing of water called a seiche. After the wind has made the water steepen to form a wave, another force attempts to restore the water to its original horizontal state. Very small waves, those less than 1.73 centimeters (less than 1 inch) in wavelength, are called ripples, or, in technical terms, capillary waves, because their restoring force is the same force that causes capillary action: the tendency for water particles to organize themselves with the minimum surface area. Gravity is the restoring force for most wind-generated waves, which are accordingly termed gravity waves. As the crests of the waves are pulled down by the force of gravity, momentum carries the water beyond the flat water position to become a trough. The waves continue their up-and-down movement as long as the wind is blowing or, if the wind stops, until the energy is dissipated and the water returns to its original position. Waves can be compared to the bungy jumper who jumps and keeps moving up and down as the energy is drained from the bungy cord until the cord reaches an equilibrium position and motion stops.

We describe the speed of a wave's motion by the period of time it takes for one wavelength, measured from crest to crest, to pass a reference mark, such as a post on a wharf or a float in the open water. This interval is called the wave period. The periods of normal waves on the sea surface range from a few seconds to about 15 seconds (s). Waves—especially small ones like light waves or radio waves—are also measured in frequency, the number of wavelengths that pass a reference point in 1 s. For example a 10-s wave would have a frequency of 10^{-1}.

The physical transfer of energy from wind to wave is complicated and not completely understood. The size of the wave itself, however, increases with the speed of the wind, its duration, and the extent of the open water over which it travels—this distance is called the fetch. The frictional drag of a slight breeze over a calm sea initially generates a series of ripples, which can also be described as waves with a period of less than 1 s. As soon as the wind has a surface against which to push—the ripple—it transfers more energy to this surface, and the waves begin to increase in size. The smaller waves build into higher, longer period waves. When the wind has been blowing hard for many hours and has a long fetch, the waves become large and powerful. These waves, which are directly under the influence of winds that generate them, are referred to collectively as sea.

When different wave patterns caused by varying wind strength and direction come together, they form a complicated sea surface, drawn here based on photographs of the surface. A photograph of such a sea state is shown below the diagram.

Because waves are so variable in time and space, families of waves of different sizes, moving in different directions, are all superimposed at any given location. These complicated sea states are analyzed and recorded by oceanographers to establish the significant wave height: the average height of the highest one-third of the waves that form in the time period being analyzed. This parameter is used as an index of wave energy in studies of wave climate.

PROCESSES THAT SHAPE THE COAST

THE MOTION OF WATER IN A WAVE

We have said earlier that the water itself does not advance with the movement of the wave. This statement seems to contradict what we see before our eyes. But think of a fishing fleet or a ball bobbing up and down in the same place while a wave passes by, or even visualize a boat resting on the sea surface while the waves march along beneath it. If the water were moving with the waves, these floating objects would move with the waves. But they do not—because the water actually moves in a circular, or orbital, path. The water particles within the wave are moving forward on the crest and backward on the trough, with vertical motion occurring between the two. When we float out beyond the surf, we can feel our bodies moving in this path; downward and away from shore, upward and toward shore. The orbital path in which the water particles move has a diameter that is equivalent to the height of the wave. Below the surface of the wave, the orbital paths decrease in diameter, becoming ever smaller until there is no motion at a depth that is equal to one-half the wavelength. Everyone who has done scuba diving knows that when surface waves are present, conditions are rather uncomfortable near the surface, but the conditions improve with depth as increasingly calmer water is reached.

Actually, there is a slight net forward motion of water in surface waves as a result of friction between the wind and the water surface. Nevertheless, water running to the shore returns to the sea because of gravity.

Water particles in a progressive gravity wave move in an orbital pattern. Their motion decreases with increasing depth. The small inset shows that water particles in fact move not only in an orbital path but also progress slightly forward in the direction of the wave.

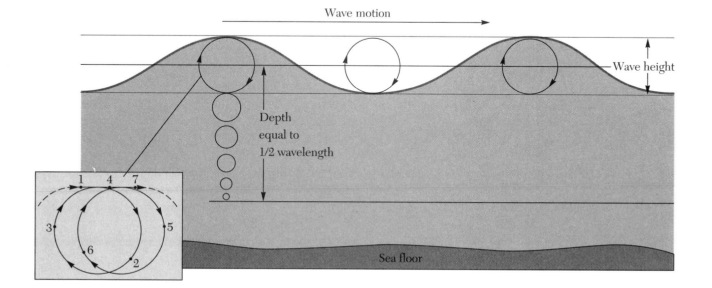

As wind waves approach land they begin to break at different distances from the shore, with larger waves engulfing the smaller ones.

Wind Waves and Swell

Wind-propagated waves often travel well beyond the range of the wind, or the wind may die down and the waves continue after and beyond the place where they were formed. During the time that a gravity wave is directly under the influence of the wind, it is called a wind wave. Wind waves tend to have a peaked crest and a broad trough, because the crest is where wind friction with the water surface is greatest. In fact, the crest often develops a whitecap as the peak is blown off by the wind. As the wind velocity increases and waves absorb the wind's energy, the waves grow higher and their periods become longer. These waves are often called storm waves. When wind waves have a wide range of heights, their pattern is complex, or choppy; the larger waves override the smaller ones, which then become superimposed on the surface of the larger ones. In the open ocean, wind waves are generally 50 to 100 m in length. Storm waves may be more than twice that long.

Occasionally a dangerous "rogue" wave appears on the open seas, up to as much as 40 m in height. A rogue wave seems to develop quickly from a pileup of small, superimposed storm waves. It exists only for a short time and is essentially solitary.

When waves move beyond the influence of the winds within a local storm or when the wind stops, the waves that persist are called swell. Swell waves have the same origin as wind waves, but after traveling a long distance they have outrun all of the smaller period waves generated with them. The long swell waves have a low trough and a small height, and they lack the steepness of the wind-driven waves. The waveform of swell is different from that of sea because the winds are no longer pressing against each wave and the friction causing the typical steep crests is absent. In profile, swell has a symmetrical, undulating, sinuous form. Although a wind-driven storm wave can be as long as 250 m from crest to crest, swells twice that length are not uncommon, especially in the Pacific Ocean, where they can build up over a greater distance than in the Atlantic or Indian ocean.

Swell, like the wind-driven wave, is a gravity wave, and its propagation is due to that force. Without anything to stop or slow the progress of these gravity waves, they can move across the entire Pacific Ocean. There is considerable power in these waves, especially when they have been developed from storm waves, because they do not peak in deep water and so their energy is not lost there. It is these swell waves that produce the huge surf in Hawaii and to a lesser extent in southern California.

APPROACHING THE SHORE

In deep waters the motion of a surface wave is not influenced by the topography of the ocean floor. As the wave moves into increasingly shallow waters near the coast, however, it begins to "feel" bottom—the orbital path of the water within the wave comes into contact with the seafloor below. The bottom friction makes the wave travel slower, so the wavelength decreases and the wave steepness increases as the waveforms are squeezed

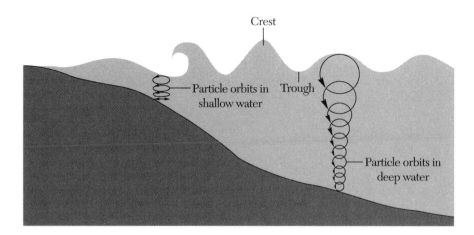

The effect of a shallowing bottom near the coast is to slow and steepen the waves and to cause the circular orbits to flatten and become smaller. The diagram shows considerable vertical displacement.

together like the folds of an accordion. The wave period, however, remains the same. This friction and interference from the bottom cause the circular path of each water molecule to be squeezed into an ellipse. Eventually the path is reduced to a simple back-and-forth motion.

Bottom friction also causes the forward motion of the wave to slow more at the bottom than at the surface. The combination of its increased steepness and differential forward motion causes the wave to become unstable. The wave crest topples forward and collapses; this phenomenon is called breaking. The wave does not lose all of its energy at this point, so it re-forms as a smaller wave, which may break again before it reaches the shoreline and loses all of its energy.

Wind waves and swells break differently as they approach the shore. The large, low swell waves begin to feel the bottom in relatively deep water, because of their long wavelength. Coming up the slope to shore, the swell wave slows gradually, gaining height and steepness until it breaks with a large curling motion and a sudden loss of energy. This dying wave is the plunging breaker. It is the classic wave beloved by surfers.

The other common breaking wave is called the spilling breaker; it is produced by wind waves as they enter shallow water. The wave breaking on the shore behaves like water spilling out of a container. The wave gives out over a period of several seconds and over a long horizontal stretch of shoreline. The energy released by an average spilling wave 1.2 m high along a 1.6 km (1 mile) of shoreline is said to be sufficient to light up 10,000 100-watt light bulbs. The breaker may form a few meters from shore or, if the wave is large and the slope very gradual, it may form a kilometer or more out from shore and roll in slowly. Such breakers are typically not so large and spectacular as breaking swell, but a surfer can ride for a very long time as these waves slowly dissipate in their shoreward progress.

Swell waves that break on steep beaches form collapsing breakers that are hard to distinguish from spilling breakers. The wave begins as a plunging wave but steepens so quickly that the leading face of the wave gives up its energy by collapsing before it can form a curl, the typical shape of a plunging wave. The wind wave on a steep beach, however, changes from a spilling breaker to a surging breaker. Just as the wind wave begins to break, it steepens instead of spilling and runs up the beach, dissipating its energy along the way.

As approaching waves encounter the features of the shoreline— sandbars, rocky outcrops, cliffs, breakwaters—they respond much the way light rays do. They reflect, diffract, and refract. Ocean waves reflect from the surface of an obstacle. They spread into the water behind projecting

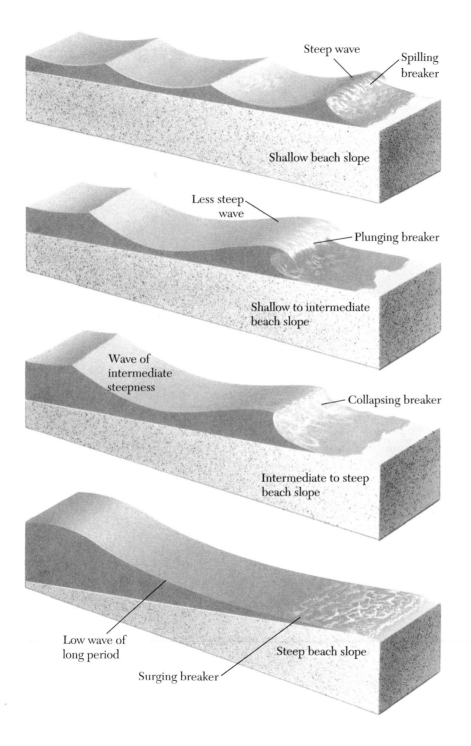

Steep wave

Spilling breaker

Shallow beach slope

Less steep wave

Plunging breaker

Shallow to intermediate beach slope

Wave of intermediate steepness

Collapsing breaker

Intermediate to steep beach slope

Low wave of long period

Steep beach slope

Surging breaker

Various types of breakers may develop in the surf zone, each caused by a different combination of wave type and nearshore slope.

forms the way light spills into the shadow of a pillar. They are bent by obstacles in the same way that light is bent as it travels through a camera lens.

Reflection

When an ocean wave approaches a cliff or seawall rising straight up from deep water, it is reflected and retains almost 100 percent of its energy—like a light ray reflected from a mirror. The force of the wave at the place of contact with the vertical wall is apparently benign, but continued wave impact will eventually lead to fatigue and failure of the wall.

When a wave approaches a cliff or wall at an angle, the angle of reflection will be equal to the angle of incidence, that is, the angle at which the wave strikes the surface. Breakwaters are structures that are built in deep water to protect ships in a harbor. These man-made obstacles reflect waves back to the sea before they break. In most natural situations, however, a

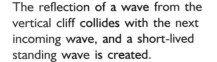

The reflection of a wave from the vertical cliff collides with the next incoming wave, and a short-lived standing wave is created.

PROCESSES THAT SHAPE THE COAST

portion of the wave energy is reflected and the rest is dissipated—partly from the natural shallowing of the water and partly from the rough surface of the obstacle which absorbs some wave energy.

The amount of reflection of energy is proportional to the steepness of the obstruction that the wave encounters. The steeper the beach, bluff, or man-made structure, the greater the amount of energy reflected. As noted earlier, a steep cliff or seawall reflects nearly 100 percent of the wave energy, whereas the usual reflection of incident energy at a beach is less than 20 percent.

Diffraction

When the path of a series of ocean waves takes it past a steep-sided island, or spit, or past the end of a breakwater, some of the wave energy will spread behind the obstacle. The diffracted energy takes the form of lower waves propagated sideways, so boats moored behind a breakwater or on the

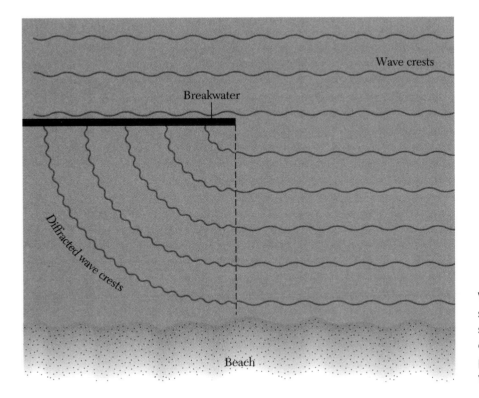

Wave energy is diffracted or spread as wave crests pass an off-shore breakwater. As a consequence, some wave energy is produced landward of the breakwater in the protected area.

lee of an island are still jostled by passing wave action. The safety of these boats during a storm depends on the height and energy of the passing waves. Sometimes the harbormaster will advise owners to move their boats farther away from the end of the breakwater to remove them from the path of the passing waves.

Refraction

A gravity wave rarely approaches the coast with its crest parallel to the shoreline; typically a wave approaches at some acute angle. As the wave enters shallow water, the parts of the wave front that reach shallow water first are slowed down, and the parts still in deeper water continue to move rapidly. Slowing thus takes place at different times along the front of the wave, because each segment of the wave slows in response to the contour of the bottom it encounters. The result is a bending, or refraction, of the wave as it passes through shallow water on its way to the shoreline. This wave refraction tends to line up each portion of the wave parallel to the shore—most of the time, however, it does not quite succeed.

As a result of refraction, a wave entering an embayment spreads into the shape of the embayment. Conversely, a wave converges at a headland jutting into the sea. Consequently, wave energy is dissipated at the shoreline in the embayment and concentrated at the headlands. Coastal scientists and engineers have constructed refraction patterns to visualize this phenomenon and the related distribution of energy along the coast. When lines are drawn perpendicular to each of the wave crests as the waves bend in response to the effects of the bottom contour, these lines—known as orthogonals because of their perpendicular orientation—show the expected spreading pattern in the embayments and convergent pattern at the headlands.

Tsunamis: Seismic Waves

Another kind of ocean wave—the most destructive of all—receives its energy from large-scale events within and on the oceanic crust. Such seismic events include earthquakes, volcanic explosions, and major coastal and submarine landslides. Popularly called a tidal wave, although it has nothing to do with tides, it is more correctly termed a seismic sea wave or tsunami—Japanese for "harbor wave," a name identifying the location where these

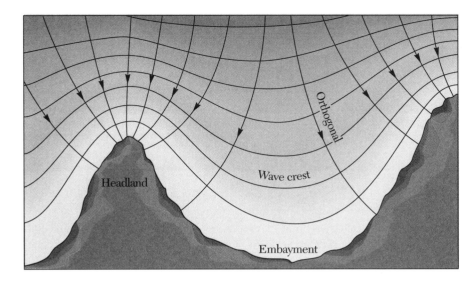

Converging orthogonals illustrate how wave energy is concentrated at headlands, and diverging orthogonals show how wave energy is dissipated in coastal embayments. The flux of energy between two adjacent orthogonals is constant.

The tsunami that struck near Minehaha, Japan, in 1983 lifted large fishing boats about 6 to 8 m above sea level.

waves are most feared. Wherever a tsunami hits, harbors are badly damaged. Many boats are wrecked, and adjacent buildings are leveled. Some tsunami landings in populated areas also cause catastrophic loss of life.

A tsunami originates when a forceful earthquake or landslide suddenly shifts or displaces a large amount of seawater and sets a train of waves in motion on the sea surface. A tsunami travels away from the point of disruption in simple progressive oscillations, much like waves generated by a stone tossed into a lake. It moves at great speeds, up to 800 km/hr, and the wavelength of each wave extends until the crests are 150 km or more apart. In deep water a tsunami often passes unnoticed because its wave height is typically only a half-meter or so. As it approaches the continental coast, it begins to slow and steepen in relatively deep water (at depths of hundreds of meters) because of its enormous wavelength. This steepening can begin as far out as 50 km offshore, and the wave that finally hits the coast is huge and energetic, as high as 25 m—a prescription for disaster. The most life-destroying tsunami on record appears to have hit Awa, Japan, in 1703. It killed more than 100,000 people.

Among the uncounted victims of the Lisbon earthquake of 1755 were those drowned by tsunami waves in Lisbon and in nearby coastal villages of Portugal and Spain. The trough of the tsunami arrived first, drawing water out of the bay and exposing the seafloor. (Recall that water in a wave moves backward in the trough.) Among the drowned were those who came to see the strange sight of the receding waters and were swept away when the crest arrived. Unfortunately, the same situation has recurred numerous times in association with this phenomenon.

In 1883 undersea explosions during the spectacular volcanic eruption of Krakatoa in the East Indies (now Indonesia) set off tsunami waves 40 m high. They swept over the nearby islands and carried a large ship 3 km inland. The known toll in lives was 36,000, but many more may have perished on small nearby islands.

In 1946 an underwater earthquake in the Aleutian Islands generated tsunamis that spread throughout the Pacific. Nearby, 30-m waves hit the Alaskan coast; only 5 hr later a 15-m wave demolished the port of Hilo in the Hawaiian Islands. By next day, waves had destroyed island villages 8,000 km distant from the point of origin in Alaska.

After the 1946 tsunami, seismologists began work on a seismic sea-wave warning system. By the early 1960s, a network of seismic monitoring stations covered the entire Pacific Ocean, the only basin where strong earthquakes are common. Knowing the location of the earthquake, seis-

PROCESSES THAT SHAPE THE COAST

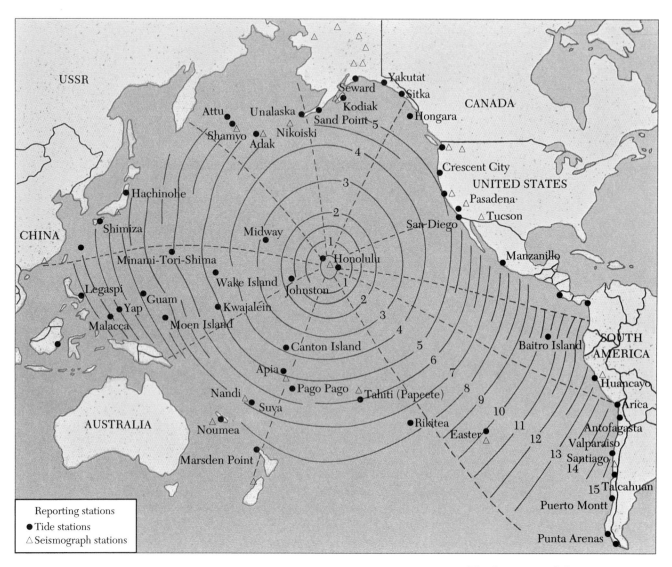

The locations of the many stations that participate in the tsunami warning system. The concentric lines indicate the travel time in hours needed for a tsunami originating from a hypothetical earthquake at Honolulu to reach locations throughout the Pacific.

mologists can now predict the path and rate of tsunami movement and provide warnings for most areas, thereby allowing at least a few hours of preparation time before the waves hit a given coast. Generally this is enough time to evacuate people. Although coasts near the origin of the earthquake may receive as little as 10 to 15 min advance notice, loss of life has been greatly reduced since the system went into effect.

STANDING WAVES AND SEICHES

In all the wave types considered so far, the waveform moves forward. Sometimes, in a constricted water body—a lake, bay, or harbor—a standing wave is produced. Such a wave oscillates but does not advance; it sloshes back and forth as it reflects off opposite ends of the basin, much like coffee in a commuter's container as the car bounces over the pavement. In this type of wave, the wavelength is equivalent to the diameter or length of the water body in which the wave develops. The form of the standing wave is an alternating up-and-down motion at each end and a node (point of no movement) in the middle. Standing waves are common in nature but little noticed because of their low height and long wavelength.

Most conspicuous of the standing waves is the seiche (pronounced "saysh"), which occurs when an external force initiates a rhythmic enhancement of the standing wave's oscillation, much like the aforementioned coffee container. A seiche is usually set up by a rapid change in weather and lasts from a few minutes to a few hours. Atmospheric pressure oscillations may occur in synchrony with the standing water oscillations and set up a resonance effect. Seiches may also be initiated when a pileup of water, such as a storm surge, at one side of a basin is suddenly released by an abrupt drop in the water or by a rise in barometric pressure. Tsunamis sometimes trap water in a bay; when a tsunami passes, the sloshing may continue for a day or two.

Seiches can cause havoc at a harbor by setting up reversing currents at the entrance or by rocking ships free from their moorings. They can also abruptly surge onto piers and beaches and sweep people away. The Great Lakes of North America and some of the large lakes in Switzerland are especially susceptible to seiches, because they are enclosed basins with large fetches and strong winds. Seiche waves have swept away sunbathers along the beaches of Lake Michigan. But the lake most subject to seiches is

The sloshing of two waves back and forth in opposite directions in a container such as a coffee cup produces a system in which the waves move up and down without propagating.

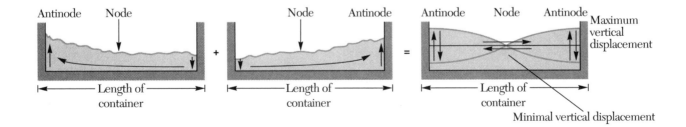

Lake Erie, with its shallow waters and orientation in the dominant wind direction. Its seiches have raised the water level more than 2 m on the eastern end while the western end near Toledo was lowered an equal amount.

Longshore and Rip Currents

When waves travel shoreward through shallow water, the water particles in the wave are displaced slightly forward, and a slow but recognizable shoreward movement of the water itself occurs as well. If the incoming waves are refracted and hit the shoreline at an angle, some of the water moves back out to deeper water and some of it moves parallel to the shore. The quantity of water in the parallel component depends on the size of the breaking wave and the angle at which it approaches the shoreline—in general, the larger the angle, the faster the water moves alongshore. As the rest of the water recedes, this component travels along the shore between the shoreline and the breaker zone of the waves. Called the longshore current, it acts like a shallow river channel, with the shoreline as one bank and the breaker zone the other. A typical longshore current moves at 10 to 20 cm/s, but a strong wind can drive a longshore component to speeds of more than 1 m/s (100 cm/s).

Longshore currents work in tandem with waves to move large volumes of sand. The orbital and back-and-forth motions of water in the incoming waves pick up the sand from the bottom as the waves steepen, and the longshore current then gathers the suspended sand from the surf and carries it along the shoreline. Because the waves approach the coast at different angles, depending on the direction of the wind driving them, the longshore current flows in one direction or the other at different times. As a result, much of the sand is simply transported back and forth along the same stretch of beach. But generally, over time, there is a net movement in one direction along the shore. And sometimes the sediment is trapped where the longshore current meets an obstacle such as a headland or jetty.

Longshore currents and sand transport (also called littoral drift because the transport takes place in the littoral zone) occur in any coastal environment where waves are refracted as they approach the shoreline. Their long-term actions can even reshape the shoreline of large bays, estuaries, and lakes. Nevertheless, by far the greatest volumes of sand are transported on open ocean beaches, where massive sandbars, spits, and other sediment accumulations attest to the carrying power of the longshore currents.

The regular movement of waves toward the shoreline with a more or less parallel orientation sometimes causes water to build up against the beach. This "pile" of water along the shore is partially supported by sandbars in the zone of breaking waves. The trapped water continues to rise slightly, wave by wave, until it is higher than the water level seaward of the surf zone. This unstable condition is called set up and is most pronounced when strong onshore winds are blowing.

The trapped water is always seeking a path of least resistance along which to return to the water level beyond the surf zone. Typically, a low area on the bottom or a break in a sandbar allows the water to move seaward until the set up is relieved. The rapid flow of this narrow stream of water is called a rip current—sometimes wrongly called a rip tide or an undertow.

The narrow rip current rushing through a saddle in a sandbar or along a low point on the nearshore bottom may reach a velocity of up to 1 m/s and extend offshore as much as 600 m. It may be difficult for the swimmer with an untrained eye to recognize, but a lifeguard or other person familiar with the shore will know the pattern of its movement. Low spots in the approaching breakers in the surf zone and clouds of suspended sediment moving seaward are the best clues to its presence. Swimmers caught in a rip current can be carried offshore to deep water. But these currents are typically quite narrow, so the safest course of action is to swim parallel to the beach, across the current rather than against it. If the current is very strong, however, the swimmer should swim diagonally away from the beach.

A rip current forms one side of a cell-like system of circulating water within the breaker zone. Water moves toward the shore in the incoming waves; after the waves break, the water flows down the shore in longshore currents, then back out again in rip currents. After moving back out through the breaker zone, the rip current becomes more diffuse and spreads out. At this point the water is caught up in the general flow toward the beach. When the incoming waves are essentially parallel to the shoreline, the longshore currents feeding the rip currents come equally from each direction and a nearly symmetrical circulation cell is produced. When waves approach the shore at an angle, the feeder currents and cells are asymmetrical.

Rip currents tend to be regularly spaced along the shore in some areas. Their size, orientation, and spacing are related to both wave conditions and nearshore bottom conditions. Along smooth coasts without sandbars, rip current spacing is typically tied to the wave conditions. In these areas rips remain at a given location for several weeks at a time, focused at the location of the smallest incoming waves.

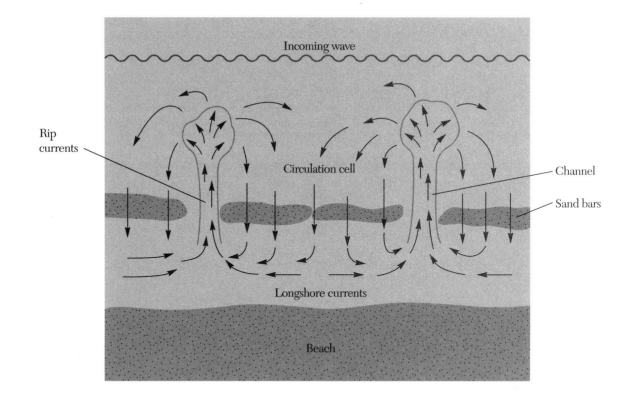

Rip currents · Incoming wave · Circulation cell · Channel · Sand bars · Longshore currents · Beach

TIDES

Along all the coasts of the world, the water level rises and falls with a regular rhythm known as the tide (Old English for "time"). Associated with this rise and fall is the tidal current—the inflow or outflow of water. Together these periodic movements of water are known as the tides. The water level rises on the flood tide until high tide is reached; then the level falls during ebb tide to the low tide level.

Sir Isaac Newton's equation for the law of universal gravitation states that all objects are attracted to one another by a force that is proportional to the masses of the two bodies and is inversely related to the square of the distance between them. This law can be stated mathematically as

$$F = Gm_1m_2/R^2$$

Rip currents develop in shallow, nearshore areas as the result of water piling up landward of sandbars, then returning seaward through low spots in the bars.

where F is the gravitational force, m_1 and m_2 are the masses of the two objects, G is the gravitational constant and R is the distance between them. Although this relationship exists for all objects, it only becomes important when the objects involved are extremely large—for example, celestial bodies—and not very far apart.

For the Earth, the other nearby celestial bodies are the Sun and the Moon. The Sun is huge but very far away, whereas the Moon is much smaller but also much closer. Although the Sun has a mass that is 27 million times that of the Moon, it is nearly 150 million km (93 million miles) away; the Moon is only about 400,000 km (247,000 miles) away. On the basis of these values, we calculate that the gravitational attraction between the Moon and the Earth is just about twice that between the Sun and the Earth. During some times of the lunar month—new moon and full moon—the Moon and the Sun are aligned, and the two attractions reinforce each other. At the first and third quarter stages of the Moon, however, the Moon and Sun are at right angles to each other relative to the Earth, so their respective attractions interfere with each other.

Although there are very small distortions of the solid part of the Earth in response to these attractions, the most dramatic effect is the distortion of the Earth's liquid envelope—the oceans. This distortion is the tide, which is actually a forced wave with a wavelength of one-half the circumference of the earth. As the Moon and Earth revolve in the Earth–Moon system, the Moon pulls on this water envelope with a force that varies over the Earth's surface, depending on the actual distance between the Moon and any specific location on the Earth. The location on the Earth that is closest to the Moon, in the low to mid-latitudes depending on the season, receives a significantly greater pull than locations near the poles. The opposite side of the Earth is subjected to the least pull because it is the part of the Earth farthest from the Moon. These differences in pull produce a bulge toward the Moon on the side of the Earth that faces the Moon.

There is also a bulge in the water envelope on the opposite side of the Earth. The Earth and Moon rotate around a common center of mass, which, because of the Earth's huge mass in comparison to that of the Moon, lies within the Earth, 4500 km from its center. The motion of the Earth as it revolves around the center of mass of the Earth–Moon system creates an outward-directed centrifugal force, which is the same over the entire Earth's surface.

Thus, there are two forces operating on the water envelope around the Earth: centrifugal force and gravity. The centrifugal force is weaker than the Moon's attractive force on the side of the Earth that faces the Moon,

PROCESSES THAT SHAPE THE COAST

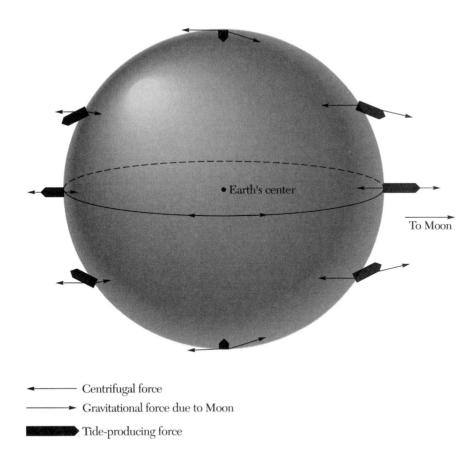

The force acting on water to produce ocean tides is the sum of the gravitational force due to the Moon and the centrifugal force due to the Earth's revolution about the center of the Earth–Moon system. (The forces are not drawn to scale.) The centrifugal force is the same over the entire Earth system, while the gravitational force exerted by the Moon varies in magnitude and direction (the angles have been exaggerated for clarity).

• Earth's center

To Moon

←————— Centrifugal force

————→ Gravitational force due to Moon

▬▬▬▷ Tide-producing force

but, as we have noted, the Moon's attractive force is sufficient to produce a bulge on that side. On the opposite side, the centrifugal force is greater than the force exerted by the Moon, so it produces another bulge because it also pulls the water away from the Earth's surface.

The bulges always retain their orientations toward and away from the Moon, but the Earth rotates beneath them. The result is that in a day—one revolution of the Earth—two bulges pass any given point on the Earth. Between these bulges are the two depressions where the water has been pulled away to form the bulges. If we look at the Earth as a revolving sphere with an elliptical envelope of water oriented toward and away from the Moon, we can see that the tidal bulges are the two high tides and the depressions the two low tides. We also see a long wave curving around the Earth, one with a wavelength equal to one-half the circumference of the

Because the Earth's tidal bulges are usually on a diagonal, the two daily tides have different ranges. That is, one high tide is higher than the other during each complete rotation of the Earth.

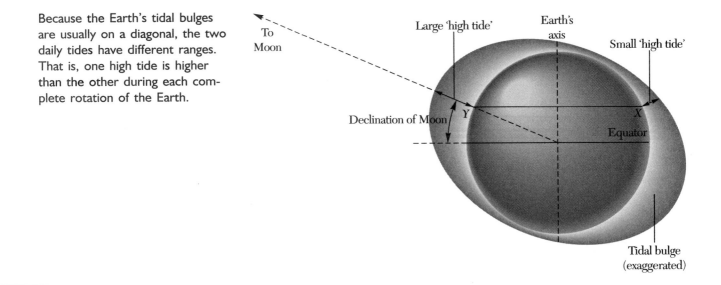

Earth. High tide is the top of the bulge, or crest of the wave. Low tide is the trough. A common component of every tidal range is the height of the tide wave, measured vertically from trough to crest (low to high tide).

THE TIDAL CYCLES

The Moon revolves around the Earth once every 27 days and 8 hours. Because the Moon is traveling in the same direction in which the Earth moves in its 24-hr rotation around its own axis, the Earth must rotate 50 min longer during each rotation for any point on Earth to arrive again at the same place relative to the Moon. This rotation interval is the length of the lunar day, or diurnal (daily) cycle of the tides—24 hr and 50 min. At any single location on Earth, the interval between the arrivals of the first and the second tidal bulges is the length of the semidiurnal (twice-daily) cycle. Each day, one high tide follows the other every 12 hr and 25 min, and each high to low tide portion of the cycle takes 6 hr and 12 min.

Of course, these are only ideal intervals, because they are based on the theoretical condition of a smooth Earth with a uniform water depth and no landmasses. The actual intervals between tides vary from place to place, day to day, and season to season. When the Moon's orbit takes it around the equator, the two daily cycles are about equal. These are the equatorial tides. More often, however, the Moon is located north or south of the equator during its monthly trip. One tidal bulge is now in each hemisphere. But the Earth rotates in the plane of the equator, so the tidal bulges are on a diagonal. Therefore, the two daily tides have different ranges. This condition results in the typical situation of unequal, semidiurnal tides at most locations, an inequality most pronounced at the tropic of Cancer and the tropic of Capricorn.

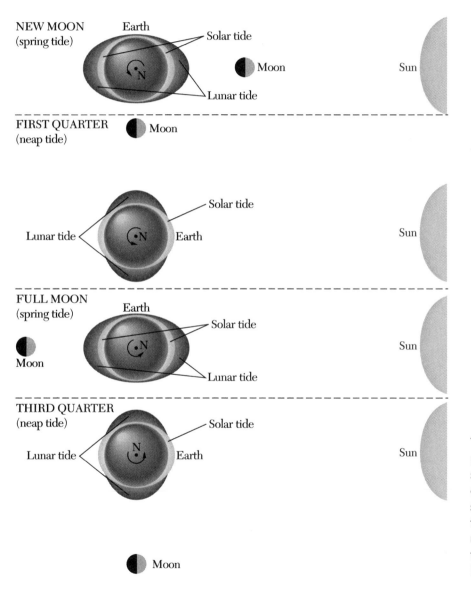

The tides at the four major phases of the lunar cycle. The solar and lunar tidal components combine to produce the large spring tidal ranges at the new and full Moons; the smaller neap tidal ranges take place at the first and third quarters when the Sun and Moon are least aligned.

The orbital path of the Moon around the Earth is elliptical rather than circular. As a result, the distance between the Moon and the Earth ranges from 221,463 to 252,710 miles as the Moon proceeds along its orbital path. When the Moon moves closest to Earth, we have exceptionally high tides, called perigee tides. When the Moon is farthest away, at apogee, high tides are exceptionally low and the tidal range is smaller than during perigee tides.

The position of the Sun in relation to the Moon has a large effect on the tidal range. Twice a month, during the new moon and full moon phases, the Sun and Moon are lined up with the Earth. At those times, their combined gravitational pulls produce the maximum tidal range of the lunar cycle—the spring tides (the term *spring* here is used in the sense of gushing water rather than the season). When the moon is in its first and third quarters, its attractive force is diminished by that of the Sun, so the tidal range reaches the lowest point in the lunar cycle—the neap tides. Both neap and spring tides have a fortnightly cycle.

The Traveling Tides

The continents form barriers to the tide wave in its passage around the Earth, so, with the exception of the Southern Ocean around the Antarctic, the ocean basins are limited in their longitudinal extent. A further complicating factor in the movement of the tide waves is the Coriolis effect, which deflects moving matter on the spinning Earth in the same way that a person walking inward or outward on a merry-go-round is deflected from his or her course. As the tide wave travels from east to west with the rotating Earth, north of the equator it is deflected to the right (clockwise) and south of the equator to the left (counterclockwise).

In some ocean basins, the tide wave moves in a great circle about a point, or node, the way a wheel turns around a hub. The height of the tide wave at the hub is zero, and the tidal range increases outward. The circles are called amphidromes, after the Greek custom (*amphidromia*) of carrying a newly christened baby up to each person in a circle of relatives. The existence of these amphidromic systems explains why the tidal cycle is a bit out of phase along a given reach of coast, seeming to come down from the north or up from the south.

In the Southern Ocean below the continents, the tide moves around the world as a single diurnal wave. In the ocean basins between continents

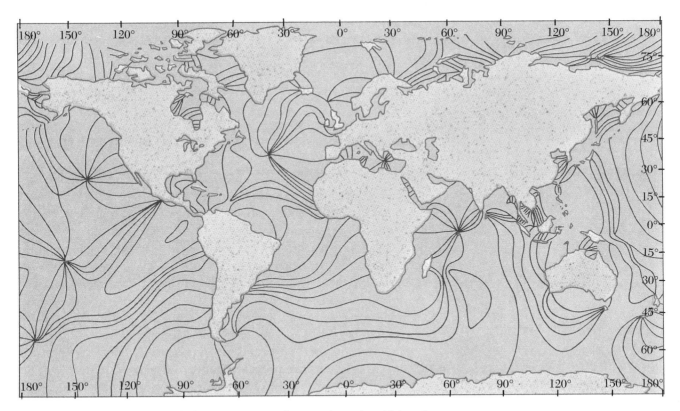

The configuration of the amphidromic systems that produce the tidal cycles.

and in the marginal seas and gulfs, the two tide waves each act as a standing wave, sloshing between coasts, with high tide at one coast and low tide at the other. More complications are added by the presence in the global ocean of all the oddly shaped landmasses, islands, land bridges, and other irregularities. Although the two global tide waves continue to move, and the sea level to rise and fall in an orderly progression under the Moon, the time of arrival and the tidal range of the wave vary from place to place. Each encounter is greatly influenced by the width of the continental shelf, which slows and steepens the wave as it approaches, and by the shape of the shore on its arrival. Some broad embayments in the coast result in amplification of the tides as water piles up during flood and escapes during ebb conditions. As a result, each location on the Earth has its own tidal character.

Tidal Range

The average tidal range over the open ocean is about 0.5 m. Each tidal bulge moves through the ocean at the rate of the rotation of the Earth, an average of about 700 km/s—the cruising speed of a commercial jet airplane. This rate varies slightly from latitude to latitude. At the continental margin, the tide wave slows by about 75 percent and steepens. A gradual shallowing and concomitant wave steepening continues up the continental shelf until the crest reaches the coast at roughly 10 to 20 km/h, about the speed of a bicycle. A further increase in tidal range occurs at bays and inlets, especially those with a funnel shape, which progressively constrict the wave and force it to steepen even more and increase in height.

Within the ocean basins, the tide wave is not discernible among all the other waves. The tidal range is measurable only at a vertical reference point, such as a mid-oceanic island, or by means of sophisticated instruments such as very sensitive pressure gauges that respond to small changes in the mass of the overlying water column. Not a lot of information is available on open ocean tides, because of their small range and because there is no real need to measure tides in that environment.

Because of the foregoing complications, it is difficult to predict the tidal range at any given location on the basis of a theoretical understanding of the tides. Predictions are best made by reviewing historical data collected at each specific location. To this end, networks of tide stations exist all along the coasts and on oceanic islands. Analysis of the record of local variations is superimposed on the fixed tidal cycles—diurnal and semidiurnal highs and lows, spring and neap tides, perigee, and others—and on the fixed amphidromic circulation systems. It is common for 40 to 100 variables to be taken into account in the prediction for a given location.

The critical data in predicting tides come from tide gauges. The first reliable tidal gauge was invented in 1882 by Sir William Thomson, Lord Kelvin, a Scottish physicist. A version of it is still the standard type. It consists of a float inside an open pipe attached to a pier. The top of the pipe, usually housed in a shed on the pier, extends from near the floor of the harbor or other monitored water body to well above the historically

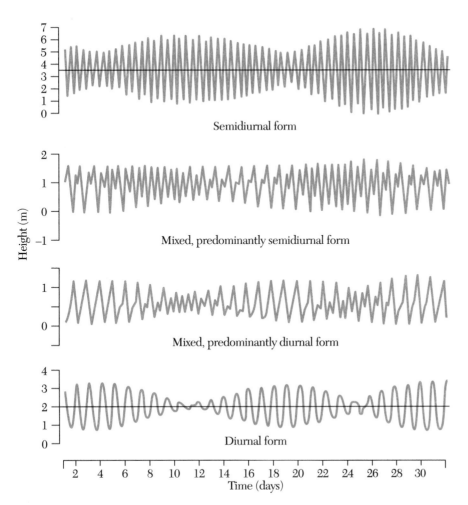

Examples of tidal curves showing each of the four major patterns that occur along the coast.

Semidiurnal form

Mixed, predominantly semidiurnal form

Mixed, predominantly diurnal form

Diurnal form

Height (m)

Time (days)

highest high tide. The base of the pipe is above the bottom or has holes in it thus preventing waves or other surface disturbances such as boat wakes from influencing its measurements; only the slow rise and fall of the tide invades the pipe. As the tide moves the float up and down, a pen records its movement on a revolving cylinder that has graph paper wrapped around it and that is driven at a constant speed by a clock mechanism. Now most stations have more modern electronic recorders that automatically transmit

High and low tides reach positions that vary over the lunar cycle: tides reach their extreme positions at the spring tide during the new and full Moon; their range is smallest at the neap tide during the Moon's first and third quarters.

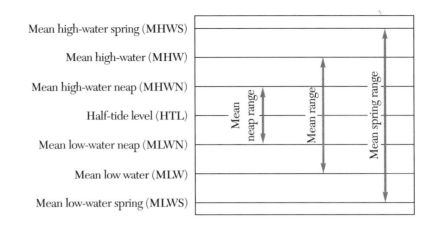

digital information to a computer tape or directly to a computer. Over a sufficiently long period of time, analysis of tidal range measurements from the tide stations enables oceanographers to determine which astronomical cycles and local variations are important at a particular location. Data are collected continuously, and the calculations are constantly updated.

Governments around the world prepare a new tide table every year for each of their major ports and harbors and for numerous other locations. The tables actually make rather thick books, which are published annually in the United States by the National Oceanographic and Atmospheric Administration (NOAA). The coastal area covered is quite extensive; for example, the volume for the east coast of the United States also covers the coastal areas from Greenland through the east coast of South America. The table lists the times of the high and low tides for each day of the year as well as data on the mean tidal ranges. The reference level used is mean sea level (MSL), which is reestablished for each location every few decades. The mean high water (MHW) datum is of particular interest to coastal residents because it serves as the boundary between private property and the public domain on the shore. It is also used as a reference point for determining when to issue special warnings for high-risk tides during storms, especially hurricanes.

Storm Surges

Storms and near-storm conditions increase the tidal range, raising the high tide level beyond the level predicted on tide tables—sometimes dangerously so. Coastal warnings are issued on the basis of the predicted level and the weather reports. The rise in water level often begins with a local atmospheric low pressure system that lifts the sea surface. High winds that rush into the low pressure area add to the rise. The meteorological energy generates storm surges, or, as they are sometimes called, wind tides or storm tides.

Depending on wind direction and coastal configuration, the surge may move toward the shore as a positive storm surge or away from the shore as a negative storm surge. Coastal residents fear the positive storm surges, which commonly cause flooding. The negative surges, however, expose areas usually underwater, and many marine organisms perish.

Storm surges as well as astronomical tides are recorded on the tide gauges. The contribution of a storm to the tidal range is measured by subtracting the predicted tide from the observed water level. The highest high astronomical tides occur when the full or new Moon is nearest the Earth, that is, when the spring tide combines with the perigee tide. The most devastating storm surges occur when an intense hurricane or winter gale coincides with a high perigean spring tide. If the storm winds arrive only a few hours later, the perigean spring tide is at its lowest low, the best possible situation for the affected coast—one reason the timing of a hurricane's progress is so important in issuing coastal warnings.

A positive storm surge flooded a San Diego beach when combined with the high tidal stage in January 1988.

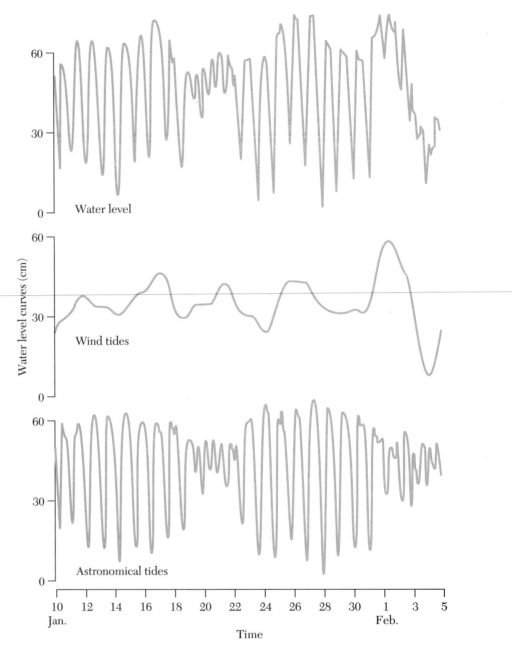

Over the course of a lunar cycle, a tidal gauge on Mustang Island, Texas, recorded the water levels shown on the top plot. The bottom two plots show the contributions of wind tides (storm surges) and astronomical tides to the total water level.

Hurricane Hugo, which hit the coast of South Carolina in September 1989, illustrates the relationship between astronomical tides and storm tides. When Hugo struck the South Carolina coast, the astronomical tide was already at its maximum of nearly 3 m. The storm surge roared onto the coast and reached up to 11 m in some of the estuaries. The consequences were devastating.

For a storm of Hugo's magnitude, a combination of energy sources coincides to contribute to the surge. First, the hurricane must be traveling slowly, because rapidly moving hurricanes do not give the storm surge time to develop to full height. Hurricane Andrew in 1992 struck the Miami area and moved through very rapidly. Although there was tremendous damage, it was predominantly due to the high-velocity winds, not the storm surge.

Second, the strong winds associated with a hurricane must be moving toward the shore (onshore winds). Then the wind-driven waves contribute to the damage by riding in on top of the surge. In the great Galveston, Texas, hurricane of 1900 that cost 5000 lives, the storm surge rose to 5 m and wind-driven waves superimposed on the surge added another 7 m to create a tsunami-sized wave.

AWFUL CALAMITY. GULF TIDAL WAVE, SEPTEMBER 8TH 1900.

An artist's conception of the flooding and destruction during the hurricane that struck Galveston, Texas, in 1900.

A third factor contributing to the destructive power of a hurricane is the configuration of the adjacent continental shelf over which the hurricane travels. A broad and gently sloping shelf facilitates a high storm surge by allowing the wind to push large amounts of water shoreward across the relatively shallow offshore waters. For this reason, the storm surges on narrow, steep coasts are not as large as those on trailing edge coasts or broad gulfs; and hurricanes on the Atlantic and Gulf coasts are generally much more devastating than those on the Pacific coast of Central America, where hurricanes occasionally strike the coast.

What happens when hurricane winds create a negative storm tide? In 1960, when Hurricane Donna swirled from the Gulf of Mexico across the Florida peninsula, the counterclockwise wind blew across Tampa Bay (on the west coast of Florida) from the landward side just at the time of low tide. As the tide surged out of the bay, the water level dropped more than a meter below normal. Natives and tourists alike came out to happily scoop up the stranded fish.

Tidal Currents

We have been speaking of the tide as a forced wave that raises and lowers the water level. The concomitant displacement of water as the tide moves between these high and low levels produces a current. Its speed approaching shore is perhaps 15 cm/s. At the shoreline, along the beach, the tidal current is nearly imperceptible and is overwhelmed by the presence of the waves, which typically dominate beach processes. There is generally a lull of about $\frac{1}{2}$ to 1 h in the current at the turning of each tide, that is, after high tide and after low tide. These short slack tide periods (so-called slack water) are attributable to the broad, flat crest and trough of each tide wave, which take time to pass a given point, and to the momentum that must be reversed from an onshore to an offshore direction for the flood cycle and the reverse for the ebb cycle.

The first coastal environment to be affected by the strong tidal current is the tidal inlet. Here large volumes of water are forced into and out of a narrow constriction during every tidal cycle. The speed at which the tidal current moves depends on the tidal prism, which is a measure of the tidal flux and is the product of the tidal range and the area of the back-barrier bays being served. Although large tidal ranges tend to generate rapid currents, many locations have small tidal ranges but large bays; this latter combination also produces a large tidal prism and, therefore, rapid currents. Speeds of a meter per second are common, with extremes of three

This satellite composite image shows Morecambe Bay, just north of Blackpool, England. Highlighted in magenta are the vast expanses of sand banks and mud flats within the bay that are covered by the sea at high tide and exposed at low tide. The tidal effects can be observed to extend several kilometers inland up the numerous river channels.

times that in some locations. Tidal currents may also be strong as the tide floods and ebbs in river mouths.

Although the pattern of water level changes is nearly symmetric as the tide floods and ebbs, the speed of the tidal current shows numerous variations with respect to time. A look at a time–velocity curve for the main channel of an inlet shows a rapid rise in velocity and then a similar decrease followed by a second rapid rise in velocity to nearly 100 cm/s during ebb conditions. But the ebb portion of the cycle takes longer than the flood. This asymmetry generally reflects the addition of fresh water from rivers that empty into the bay.

In the next two chapters, we will examine the broad delta and estuary regions of the world's coasts and will see how the local tidal currents contribute to the shaping of the coast.

4

Estuaries, Salt Marshes, and Tidal Flats

Although bays vary widely in size, shape, and origin, the most interesting and meaningful way in which they differ from one another is in the behavior of their waters. Most of them can be separated into two large categories—lagoons and estuaries—on this basis.

Lagoons are quiet saline bays with no regular freshwater influx. Generally they are protected from the open ocean by a barrier island, a reef, or an obstruction that prevents wave attack and inhibits tidal circulation. Estuaries are more turbulent bays receiving fresh water from rivers and salt water from the sea. Although the terms estuary and lagoon distinguish two quite different geological and biological environments, they are often loosely applied, even in scientific literature.

In this chapter we will look at various estuary types. We will also consider their associated environments—tidal flats, coastal salt marshes, and mangrove swamps. Lagoons will be discussed in a later chapter, along with barrier islands.

A low coastal marsh of *Spartina alterniflora* grows along an estuary margin in Carteret County, North Carolina.

The term bay refers to a large category of coastal settings—a concavity or indentation of the coastline that is smaller than a gulf but larger than a cove and that generally is somewhat protected from open-ocean wave energy. Names are not a reliable guide to the true nature of many coastline irregularities. The Bay of Bengal, for example, is one of the world's largest gulfs.

Bays vary in shape and origin. On leading edge coasts, bays are generally long and narrow. They lie above faults and fractures that have allowed ocean waters to extend inland. Along trailing edge coasts, bays are usually wider because they often cover river valleys drowned during the past several thousand years by the rising waters from melting glaciers. In northern latitudes glaciers have carved deep furrows in the continental margins. As the glaciers melt and sea level rises, ocean waters invade these furrows, creating the scenic, elongate bays called fjords. Other bays have wide mouths and appear as broad curves on the coastline. These bays form as a result of differential erosion of, for example, sandstone, a rock type easily worn down by waves, and the more resistant granite headlands. Such an embayment is carved from the sandstone and is held in place by the granite ramparts on each side.

A quiet lagoon is protected by a barrier island on the Peninsula de Quevedo, Sinaloa, Mexico.

ESTUARIES, SALT MARSHES, AND TIDAL FLATS

TIDAL RANGES AND TIDAL BORES

An estuary is an arm of the ocean that is thrust into the mouth and lower course of a river as far as the tide will take it. Every estuary has three main sections: The inland end, where the river enters, is called the head. The middle part is the fully estuarine area, where fresh water and salt water mix. The seaward end, at the indentation of the coastline where the ocean enters, is called the mouth.

Estuaries with wide mouths and narrow heads have a large tidal range because each tidal wave that approaches a given length of coastline transports a given amount of water into an increasingly narrower part of the estuary. This geometry produces an increase in the high tide level. The ebbing of the same amount of water results in a similar relative decrease in the low tide level. Tides that roll into maritime Canada's funnel-shaped Bay of Fundy increase in range from 2.4 m at the mouth on the Gulf of Maine up to 16.3 m in the Minas Basin, a relatively isolated part of the estuary located at the landward end. These are the largest tidal ranges in the world, nearly eight times those of the nearby open coast.

A kayaker rides the crest of a tidal bore near Mt. San Michel (upper left) in the Bay of St.-Malo on the Normandy coast of France.

A tidal bore flows up the Amazon River at Furo do Guajuro, Brazil.

Some rivers located at the landward end of an estuary experience another extreme tide-dependent condition—a tidal bore, an abrupt and migrating rise in the water level at the beginning of the flood tide. This wall of water is a response to the quick reversal from an ebbing tidal condition to a flooding one. Bores are uncommon, forming only in special circumstances that depend on tidal conditions and the morphology of the estuary. The bore in the Truro River of the Bay of Fundy is typically only about a half-meter high. In the Bay of St.-Malo on the northern coast of France, a bay with the world's second largest tidal range, the bore rarely exceeds a meter in height. Large tidal bores occur in the Pororoca River, a branch of the Amazon, and in the Chien-tang estuary in China. The bore reaches 5 m in the Pororoca and nearly that height in the Chien-tang.

ESTUARINE CIRCULATION

There are various ways in which the influxes of fresh water and seawater into an estuary can interact. In the simplest estuarine circulation pattern, the water masses remain separated: Fresh water flows seaward in the upper

ESTUARIES, SALT MARSHES, AND TIDAL FLATS

layer and the denser seawater flows landward in the bottom layer. As the tides change, the flow rates of fresh water into the estuary and of seawater in and out of the estuary vary systematically. Although a river discharges fresh water at a fairly uniform rate, the tides influence freshwater flow into the estuary. The flooding tide moves into the river mouth and pushes against the fresh water, thereby slowing discharge. The opposite condition occurs during ebb tide. An analogous condition persists at the mouth of the

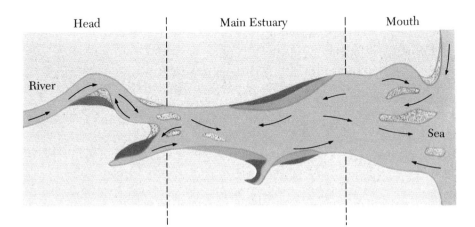

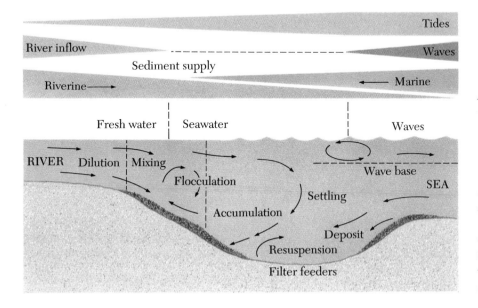

Above: A typical estuary can be divided into three major parts, based on the relative influence of important processes (*center*). River inflow dominates in the head, and waves and tides in the mouth. *Below:* River water and seawater mix as they converge. Here the incoming tide forces the outgoing river water downward. Sediment is deposited on the estuary floor when the water movement is slowed.

estuary, where the water flows into the open sea. Flood tides retard the flow and ebb tides enhance it. Therefore, tidal influences are both regular and predictable, except when tidal range varies significantly because of a storm surge.

Generally, however, circulation in an estuary is much more complicated; and much of this complexity is due to the difference in the nature and, therefore, the behavior of the two water masses—fresh water and seawater. Seawater has a salinity—the content of total dissolved solids—of about 35 parts per thousand, or 3.5 percent; fresh water has essentially zero salinity. This difference in salinities leads to different densities for the two water types: 1.000 g per milliliter (mL) for fresh water and 1.026 g/mL for seawater. Although the difference in densities is small in absolute numbers, in the absence of significant disturbance by waves or currents, it is sufficient to allow gravity to pull the heavier salt water below the level of the lighter fresh water. Stratification of the two types of estuary water—fresh water "floating" on salt water—is the simplest circulation pattern.

In the 1950s Donald Pritchard, a coastal scientist at the Chesapeake Bay Institute of Maryland, identified and described three types of estuarine circulation: stratified, partially mixed, and fully mixed. He based his classification scheme on the interactions of fresh water and salt water.

In a stratified estuary, the freshwater and saltwater masses are almost completely separate; no significant amount of mixing occurs. The upper, freshwater mass flows as a distinct layer from the head to the mouth of the estuary. The subsurface saltwater mass flows along the floor of the estuary underneath the freshwater layer, with incursions and excursions as the tides flood and ebb, respectively. The incoming saltwater mass takes the form of a wedge as it proceeds up the slope of the riverbed.

Estuaries dominated by large rivers have a stratified circulation pattern. In the Hudson River estuary, the saltwater wedge sometimes extends 100 km upriver, as far as the city of Newburgh, New York. This stratification can exist here only because there are no physical processes such as waves or strong tidal currents to mix the two water masses.

In a partially mixed estuary—Chesapeake Bay is a good example, tidal currents are the dominant factors in circulation. Although freshwater discharge from the James and other rivers is rapid, tidal currents are strong and turbulent relative to those in stratified estuaries. As a result, there is mixing at the interface between the upper freshwater layer and the lower saltwater layer. This mixing produces brackish water, that is, water with salinities that are typically 15 to 20 parts per thousand. Therefore, the

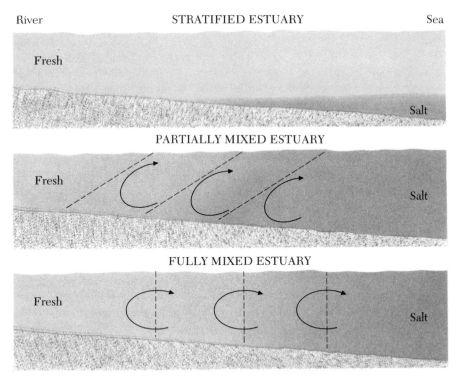

River   STRATIFIED ESTUARY   Sea

Fresh

Salt

PARTIALLY MIXED ESTUARY

Fresh

Salt

FULLY MIXED ESTUARY

Fresh

Salt

Diagrammatic representation of the three types of estuaries based on the interaction of fresh water and salt water. In nature, there is a complete transition between them. Dashed lines represent isohalines, lines of equal salinity.

salinities in a partially mixed estuary form a gradient that ranges from 0 for pure fresh water to 35 parts per thousand for undiluted seawater.

Because of partial mixing, this type of estuary contains fresh water, salt water, and water with intermediate salinities. The positions of each water type vary from estuary to estuary, and with time and conditions in any given estuary. The best way to show these positions is by plotting isohalines, or lines of equal salinity. Typically, salinities in a vertical water column in a partially mixed estuary are highest near the bottom and lowest at the surface. An isohaline in the vertical plane slopes down from the seaward side of the estuary to the river mouth, because on the surface (the horizontal plane), salinities become higher and higher in the seaward direction, until the salinity of undiluted seawater is reached. At some position in the mouth of the river, we find the zero isohaline—or fresh water. Remember that this three-dimensional distribution of salinities is very dynamic and can vary in as short a time period as a single tidal cycle. It also varies with the seasons or in relation to a storm and subsequent flood event.

The fully mixed estuary produces a homogeneous profile of water salinities; that is, at any given place within the estuary, the salinity is the same from the surface to the bottom of the water column. Consequently, isohalines in the vertical plane have no slope. In the horizontal plane, there is pure fresh water at the river mouth and fully saline seawater where the estuary opens into the ocean, with brackish water in the middle.

Several conditions can produce a fully mixed estuary. One very common situation is found in large, shallow estuaries like those sprinkled along the coast of the Gulf of Mexico. The large waves entering such a shallow bay have motion extending to the bay's bottom; therefore, they completely mix its waters from top to bottom. In other situations very strong tidal currents like those in Delaware Bay or the Bay of Fundy, or very high river discharge like that at the mouth of the Amazon can also produce this fully mixed condition. All these situations produce extreme turbulence within the estuary.

Many estuaries experience seasonal shifts in circulation patterns. A shallow estuary in which mixing depends on wave action will revert to the stratified pattern in the summertime when waves subside. A large river that produces a partially mixed estuary during the spring snowmelt, when its currents are strongest and most turbulent, will change to a stratified pattern in midsummer when its current subsides. In winter, when an estuary's surface waters are churned by storm winds and become more dense as they cool, the upper and lower layers will mix temporarily.

Other estuaries have the same circulation pattern throughout the year. For example, tidal currents in the Bay of Fundy and other funnel-shaped estuaries are strong enough to mix the water completely, regardless of the season.

SEDIMENT DEPOSITS

Coastal geologists refer to estuaries as sediment sinks because they are basins in which large quantities of sediment accumulate. In fact, the geologic lifetime of an estuary is short because of this characteristic. Some estuaries fill so rapidly that they are short lived even in historical time. When in the late thirteenth century Marco Polo visited the port of Hanghou (he called it Kinsai) on the Chien-tang estuary in northeastern China, it was a great commercial city with over a million inhabitants. Less than 200 years later, the bay had filled with sediment and trade had moved elsewhere.

River runoff and incoming tidal currents provide sediment from both the landward and the seaward directions, so estuaries tend to fill in toward the middle as they mature. If sea level rises during infilling, then the space for sediment increases. If sea level falls, then the estuary is left "high and dry." Many estuaries must be dredged periodically to keep them open for shipping. Virtually all large ports on estuaries experience this activity on a regular basis.

River-deposited sediment tends to be a mixture of sand and mud, whereas sediments brought in from the sea generally consist of sand mixed with marine shell gravel; mud is less commonly brought inland from the seaward end because outflowing currents generally carry it far out on the continental shelf. Within the estuary, mud is typically transported as suspended load and sand is carried as bed load, rolling and bouncing along the floor of the estuary. In low-energy estuaries with sluggish currents, fine-grained sediment accumulates on the bottom, although some of it is carried out of the estuary with the ebb tides.

Estuaries, again like deltas, can be divided into three types on the basis of the most energetic factor influencing sediment distribution and deposition. We can, therefore, speak of river-dominated, tide-dominated, and wave-dominated estuaries.

River-dominated Deposition

When a river brings a high sediment load to an estuary basin with weak tidal currents and small waves, the sediment accumulates rapidly, forming a bayhead delta at the river mouth. Several large estuaries on the Texas coast of the Gulf of Mexico have such deltas: San Antonio Bay fed by the Guadalupe River; Corpus Christi Bay, by the Nueces River; and Galveston Bay, by the Trinity River. In front of these deltas and that at Mobile Bay in Alabama, bars and barrier islands block the open-water waves and dampen the incoming tide to a range of much less than a meter.

Estuaries with several rivers do not normally develop bayhead deltas. Chesapeake Bay, also a sheltered bay, receives water from many rivers, but their sediment load is deposited along the banks and shores of the complex bay rather than building up at a single place to form a delta.

The factors influencing bayhead delta formation are related to the geomorphologies of these two locations. The Texas coastal plain is nearly flat, and each river empties abruptly into its own estuary. It dumps its sediment load equally abruptly, thereby forming a delta. The Chesapeake Bay rivers drain a hilly terrain, and their mouths are relatively long and narrow, so their sediment load is deposited gradually and does not accumulate fast enough to form a delta. Both the Texas coast and Chesapeake Bay estuaries, however, are river-dominated.

Tide-dominated Deposition

A tide-dominated estuary has no barrier or other construction at its mouth; rather it tends to be funnel shaped. The dampening effect that barriers have on tidal flux is not present, and the shape of the bay focuses the flood tide waves. Both conditions maximize the influence of tides; their strong currents and extreme turbulence totally mix the waters of the estuary.

The combination of large tidal range with strong tidal currents results in a sand-dominated estuary floor, because the mud either is swept out to sea or is trapped at the landward limits of the estuary. The Bay of Fundy on the western North Atlantic coast and the Gironde estuary on the northwest coast of France both have the sandy floor of a tide-dominated estuary.

Plotting the rise and fall of the tides in relation to time produces a diagram that shows the changes in water level as each tidal cycle passes; this curve is essentially symmetrical. Sediment transport is associated with the tidal cycle through the changes in velocity of the tidal currents that are produced by the rise and fall of the water level at a single location. These

Map showing that surface sediments tend to become fine along the margins and toward the upper end of Cobequid Bay in the Bay of Fundy, Nova Scotia.

Bedrock	Medium sand	Mud
Sand and gravel	Fine sand	

ESTUARIES, SALT MARSHES, AND TIDAL FLATS

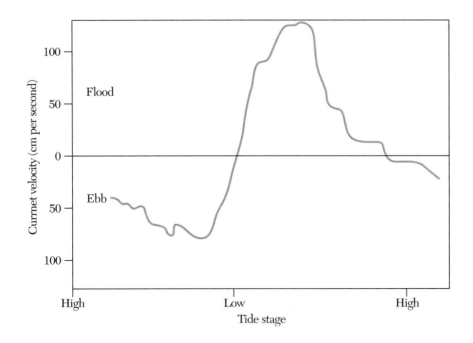

A typical time–velocity curve for an estuary shows the current velocity reaching its peak midway in flood tide. The curves are irregular from the influence of variations in the estuary floor and the overall shape of the estuary.

data are obtained by moored or hand-held current meters. From the baseline flow of the low tide condition, currents increase in speed, reach a maximum sometime near the middle of the flood cycle, and then gradually slow down until slack water at high tide. The ebb cycle repeats this general sequence, ending at low tide. The graphic representation of these data is an important tool for analyzing both the circulation of the tide-dominated estuary and its sediment transport. Such a representation shows considerable asymmetry in the distribution of velocities over each phase of the tidal cycle; it may also show that the durations of the flood and ebb portions are different—by more than an hour in many cases. This type of graphic plot is called a time–velocity curve, and the shape of the curve is referred to as a time–velocity asymmetry.

Each location within an estuary displays its own characteristic time–velocity curve because adjacent locations may be dominated by different factors. It is common, for example, for a channel to be ebb dominated and the adjacent tidal flats to be flood dominated. In addition, changes in the time–velocity curve for one location occur periodically within the lunar cycle and if channels or other current-influencing aspects of the estuary floor and tidal channels change.

Bird tracks across small ripples on a tidal flat in Olympic National Park, Washington.

Strong tidal currents not only move great volumes of sediment into the tide-dominated estuaries but also move sand back and forth over the estuary floor during each tidal cycle. Sand can be picked up and transported by currents flowing faster than 15 to 40 cm/s. Because estuarine currents flow at these or greater velocities for a few hours during each flood and ebb cycle, various bedforms—ripples, sand waves, and dunes—appear on the estuary floor.

Bedforms are regular perturbations of the sediment surface and are caused by irregularities in the flow of water over the sandy floor of an estuary. It is not understood exactly how the energy transfer from the flowing water produces ripples or sand waves, but it has been proposed that the roughness of the sediment—that is, of the grains themselves—causes turbulent flow. This turbulent flow moves the sediment and initially produces ripples, the smallest bedforms. The ripples create a second type of roughness, bed roughness. This surface feature retards flow over the bed and helps to form the boundary layer, which is the zone near the sediment floor where friction and roughness influence flow velocity—friction retards flow and roughness increases turbulence. In many shallow or intertidal parts of estuaries, this boundary layer extends throughout the entire water column.

In 1961 researchers at Colorado State University, led by Daryl Simons, developed a scheme that shows the systematic relationship between flow and the development and style of bedforms. Their discoveries were based on experiments in a large, long flume in which water of various velocities was passed over a bed of sand, and were later documented in rivers and other natural sedimentary environments. Initially sand simply moved as individual grains; but as the water velocity increased, ripples formed. Further increases in water velocity produced larger bedforms, generally called megaripples. The next stage, however, was a smoothing of the bedforms into a flat sheet of rapidly moving sand grains. This condition is referred to as a plane bed and represents the transition from subcritical flow to supercritical flow, a condition that is rare in estuaries.

A method of quantifying the relationship between current velocity and flow condition has been developed by considering both the velocity of the water mass over the sediment and the depth of the water. This relation is stated as

$$F = \frac{U}{\sqrt{gD}}$$

where U is the velocity, D is the water depth, and g is the gravitational constant. The parameter F is the dimensionless Froude number, which is a useful index of the energy imparted to the sediment floor. Subcritical flow has a Froude number below 1.0; for supercritical flow, F is greater than 1.0.

Unlike sand, a noncohesive sediment that is typically transported as bed load, mud-sized particles, especially clay-sized ones [<4 micrometers (μm)], are transported as suspended load. In the zone where fresh and salt water mix, these very small clay particles combine to form various larger particles: floccules, inorganic particles bound by electrochemical forces; aggregates, inorganic particles strongly bound by intermolecular or cohesive atomic forces; and agglomerates, inorganic particles bound together by organic matter or surface tension. Floccules are especially important in estuaries, particularly in the middle zone of partially mixed estuaries, where suspended clay particles first encounter the saline conditions that foster clumping. Flocculation generally takes place in water with salinities above 3 parts per thousand. The clay particles have highly negative surface charges that are balanced by the strong positive charges on the ions in seawater: calcium (Ca^{2+}), magnesium (Mg^{2+}), and sodium (Na^+). The neutralized clay particles attract one another electrochemically (van der

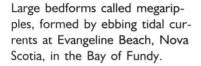

Large bedforms called megaripples, formed by ebbing tidal currents at Evangeline Beach, Nova Scotia, in the Bay of Fundy.

Waals force), and flocculation takes place. Composed of individual particles of less than 4 μm in diameter, floccules attain diameters up to 30 μm and create a characteristic region of muddy water called the zone of turbidity maximum.

The middle portion of the estuary where flocculation occurs has the most sluggish currents, because it is remote both from the river mouth and from the strong tidal currents at the estuary mouth. Consequently, large quantities of mud are deposited in the middle portion of most low-energy estuaries.

Estuaries have another form of sediment in addition to that contributed by the rivers and the sea, namely, the biogenic material produced in the estuary itself by its own population of plants and animals. Numerous organisms that find seawater inhospitable thrive in the brackish waters of the middle portion of the estuary where salinities range from 0.50 up to 17 parts per thousand. Ostracods (tiny shrimplike crustaceans in a bivalve shell), foraminifera, worms, and various snails, crustaceans, and bivalves are common estuarine animals. Shells and hard body parts from these organisms make an important contribution to the sediment of the estuary.

The physical and chemical breaking up of skeletal material produces sand and silt that is an important part of most coastal sediment deposits. Waves and currents cause breakage of shells by collision; some organisms drill holes in smaller ones, or they secrete chemical fluids that dissolve them. Shallow and intertidal rocks may be slowly eroded by snails, limpets, sponges, and sea urchins, all of which scrape the surface as they forage for food. Some clams can actually bore into rock, wood, or metal as they seek shelter and a place from which to feed.

Filter feeders, such as oysters, worms, and many bivalves, are among the estuary organisms contributing sediment. They ingest suspended sediment, extract the food particles in their gut, and excrete the rest as mud pellets. Indeed, most of the accumulated estuarine mud is actually composed of cohesive pellets that upon first glance appear to be nothing but a mass of unstructured mud. Examination under the microscope reveals that the mud is soft pellets 0.10 to 0.25 mm in diameter. By removing and reworking the fine suspended sediment that might otherwise be carried to sea, the filter feeders greatly increase the rate of sediment accumulation in the estuary itself. The pellets, too large and cohesive to be transported by anything but relatively strong tidal currents, remain on the floor and sides of the estuary and are eventually incorporated into the stratigraphic record.

Tidal Flats

More than half of the margins of most estuaries are rimmed by tidal flats—areas of mud and sand that are exposed at low tide and flooded at high tide. Their extent is determined by the shape of the estuary and by the tidal range. Obviously, a large tidal range will provide a broader intertidal environment under any given set of circumstances. Not so obvious is the influence of the gradient of the estuary margins. Some are very gradual and therefore provide wide intertidal flats. Others, however, are steep; for example, the sides of fjords or tectonically formed estuaries. These estuaries have narrow intertidal flats, even in a setting with a large tidal range. Nevertheless, much of the area of many estuaries is made up of tidal flats intersected by tidal channels. Notable in this regard are the Wadden Sea tidal estuary on the North Sea coast of northern Europe, much of the Bay of Fundy in Canada, and the Bay of St.-Malo, on the Normandy coast of France.

The same currents that distribute sediments throughout the estuary and along the shoreline also deposit them onto the tidal flats. Local waves and longshore currents play a part, but tidal currents dominate most tidal

Mt. San Michel, which rises above the intertidal sand flats in the Bay of St. Malo where spring tides reach 15 m. The parking lots in the foreground are flooded at high tide.

flat systems. Tidal flat sediment is composed mostly of mud and fine-grained sand and the shells of the small animals that have lived there; coarser grains settle out in the tidal channels. When exposed at low tide, the tidal flats have the appearance and texture of sandy mud or muddy sand.

Certain key horizons can be noted on the tidal flat surface and correspond to water levels at given tidal stages. Spring high tide occurs in conjunction with the new or full moon and is a regular and predictable inundation of the sediment surface that reaches the highest position on the tidal flat; correspondingly, spring low tide marks the lowest position. Neap tides during the first and third quarters of the moon have a smaller range and only attain intermediate positions. During a tidal cycle, the high tide location is covered by water and receives sediment for only a few minutes during slack high water. In contrast, the low tide level is covered with water and is subjected to the related waves and currents for virtually the entire cycle of 12 hours and 25 minutes. The time sequence of this cycle can be altered by weather conditions, especially by strong winds. Occasionally, positive wind tides—high water produced by strong winds pushing water up on the estuary margin—may cause flooding into the supratidal environment, the zone above the normal high tide position. The opposite situation can occur if the strong winds are blowing offshore and produce lower than normal low tides.

Particles of sediment transported onto the tidal flats follow a predictable path during a tidal cycle. The scheme was worked out after World War II in numerous detailed studies of sediment transport on the Wadden Sea tidal flats by H. Postma of the Danish Hydraulic Institute and by L. M. J. U. Van Straaten and Phillip Kuenen, colleagues from the University of Groningen in Holland. Their model, depicted in the diagram, fol-

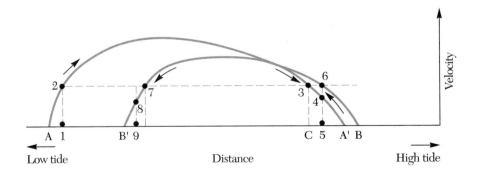

Sediment is transported up onto the tidal flat by the flood tide and back down by the ebb tide, according to this scheme. A sediment particle travels along the path represented by the dashed line as it is transported by the moving water, whose changing velocity is given by the blue curves.

ESTUARIES, SALT MARSHES, AND TIDAL FLATS

lows a particle of sediment for one tidal cycle. The particle is picked up from point 1 when the flood tide current reaches a velocity high enough to lift the sediment grain (at point 2 on the velocity curve A–A'). The current carries the particle higher onto the tidal flat along a hypothetical path shown by the dashed line. The particle begins to settle when the tidal current velocity declines to the value required to lift it (shown at point 3). The sediment particle does not drop vertically but actually reaches the bottom surface a bit further up on the tidal flat (along a path from point 3 to point 5) when the current velocity is reduced to the value at point 4. This phenomenon is known as the settling lag because a particle always moves beyond the place (C) where it begins to drop. As the tide ebbs, the particle is again picked up, when the current once again attains the same critical velocity, but this time by a water mass (shown by the velocity curve B–B') different from the one that carried it up the tidal flat. The particle begins to settle again at the same current velocity, but the ebbing water mass reaches that velocity further up on the tidal flat (shown here at point 7). The particle settles again in the same way it settled on its landward trip (from point 7 to point 9), reaching the bottom at a current velocity shown at point 8. But it comes to rest at a site further landward than that from which it began the trip several hours previously. The ebb tide gathers speed as it moves out, but because the settling lag has brought the particle closer to shore, a larger part of the ebb tide will have passed over the particle before the water reaches the required velocity to pick it up. Therefore, there is not as much time left for the particle to be carried seaward. This effect is called scour lag. This entire process of settling lag and scour lag accounts for part of the net movement of sediment particles up a tidal flat along the estuary margin, or anywhere else tidal flats occur. In addition, the ebb tide cannot carry the particle as great a distance as the flood tide can because its velocity is not as high and it does not remain at the required velocity as long.

If a tidal flat were a completely smooth and gently sloping surface, sediment would be distributed on it according to its position within the complete tidal range. The size of the sediment particle is an important variable in this scheme. Over a given tidal cycle, larger particles will be transported shorter distances than smaller ones, because of the greater mass involved. Additionally, the longer any particular location on the tidal flat is covered by water, the longer it will be subjected to currents that can move its sediment. The size of the particles and the length of time over which they are moved by currents are the fundamental variables in controlling the distribution of sediment particle size across a tidal flat. The result is that there is a regular decrease in the grain size up the slope of the tidal flat

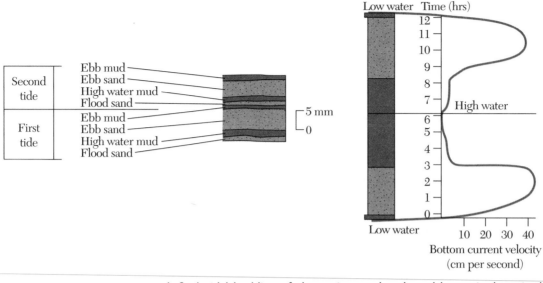

Left: A tidal bedding of alternating sand and mud layers is deposited over a tidal cycle. *Right*: Relatively coarse sand particles settle out of swift-moving water in mid-flood or mid-ebb tide, whereas the finer sediment of mud settles out of suspension only when the water velocity is low during slack water at high and low tides.

A vertical cross-section shows tidal bedding formed in the mostly muddy sediments along the margins of the Bay of St.-Malo near Mt. San Michel, France.

from the low tide level to the high tide level. Typically, low tide elevation is marked by the coarsest sediments, commonly sand, and high tide elevation is characterized by mud.

As the tide floods and ebbs, and as the grains sort themselves according to size, the sediment on the tidal flat is deposited in thin, regular layers called tidal bedding. Each individual bed or stratum in this sequence can be a few millimeters to more than a centimeter thick. The tidal cycle leaves its own imprint on this bedding, producing alternating layers of sand and mud. Two sand layers represent the flood and the ebb portions of the cycle when currents are flowing rapidly. Thinner mud layers are deposited between the sand layers at or near slack high and slack low tide, when fine sediment settles out of suspension. With the spring tides, neap tides, and storm tides also leaving their own specific record of sediment accumulation, it is sometimes possible to recognize hundreds of layers and reconstruct a coastal calendar of events. Geologists studying ancient stratigraphic records can recognize ancient tidal flats and tidal channels from their bedding characteristics and can even reconstruct the behavior of tides.

Another type of tidally produced stratification is tidal bundling, typically associated with tidal channels or relatively strong tidal currents and at least modest-sized bedforms such as megaripples. The alternation of flood and ebb tides is generally accompanied by time–velocity asymmetry. In or adjacent to channels, the ebb tide currents are strong and last longer than the flood tide currents. The additional water volume produced by the freshwater runoff into the estuary and the fact that ebb tides are essentially a gravity response overbalances the flooding tides which are a forced wave; analogous to pushing water uphill into an estuary.

The asymmetrical distribution of energy along the channel floor tends to produce pulses of migrating sands; or under more symmetrical conditions, it can cause changes in direction of the bedforms as the current reverses through the tidal cycle. As the current moves over the bedform on the channel floor, it transports sediment and thereby moves the bedform. Many people have probably seen this phenomenon on a small scale in curbside gutters or even at the beach. The dominant current, in this case during the ebb stage, moves the bedform and creates cross-stratification that is equivalent in scale to the height of the bedform. The flood stage that follows lays a thin mud drape over the steep face of the bedform, because

Tidal bundling formed by the migration of megaripples in the Oosterschelde Estuary on the Netherlands coast. The thin dark layers are muddy breaks between individual packages of sandy laminations. They were deposited during the slack water between tidal cycles. Only the dominant tides, ebb in this case, are recorded by sediment accumulation; the subordinant tides, flood in this case, did not accumulate sediment. The thin packages of sandy laminations on the upper right were deposited during the neap tide and the thick ones on the upper left during the spring tide. The horizontal straight line across the photo is an artifact of the photographic process.

spring
↓

neap
↓

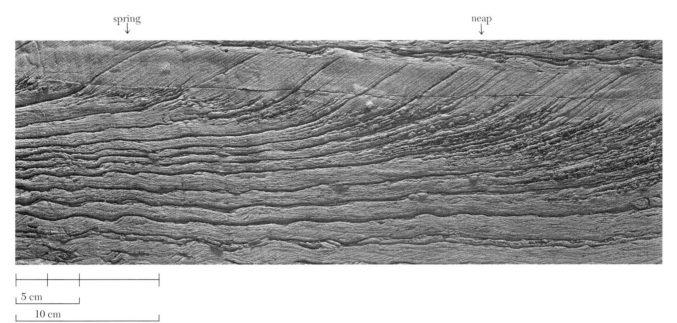

5 cm

10 cm

25 cm

these currents are not strong enough to lift and carry the sand particles and reverse the bedforms. A thin muddy seam is accumulated between each cross-stratum of sand. When viewed as a slice across the channel floor, the accumulated sediment contains a sequence of cross-strata and mud drapes that represents both the daily tidal cycle and the lunar cycle. It is common to be able to distinguish bundles of 14 cross-strata that are set apart by the relative amounts of sand and mud that accumulate; these are tidal bundles. Each bundle represents the 14-day spring-to-neap tidal cycles.

On some tidal flats, incoming waves obscure or obliterate the tidal bedding. But this destruction only happens when the energy imparted by the waves exceeds that of the tides. In other places where shallow water covers large expanses of a tidal flat for long portions of each tidal cycle, the back-and-forth motion of the waves mixes large quantities of sediment, thereby destroying any tidal signature or preventing it from being incorporated in the sediments in the first place.

The broad Wadden Sea tidal flats behind the barrier islands along the European North Sea coast offer a good example of tidal flats that have little tidal signature. These flats are flooded for 4 to 6 hr of each tidal cycle, time enough for the strong North Sea winds to generate modest waves over their surfaces. Waves produce their own small bedforms: wave-formed symmetrical ripples that produce their own style of cross-stratification. Channels and channel margins adjacent to the tidal flats still display tidal bundling, a record showing that tidal currents are dominant within the channels themselves.

Organisms that inhabit the tidal flats can also obliterate vast areas of the tidal signature or even the wave signature on tidal flats. Often the same creatures that filter the suspended sediment from tidal waters also burrow into the tidal flats for protection and for food. Clams and worms are especially active in this environment, along with a variety of small crustaceans. The burrowers turn over the sediment much as earthworms do, and they disrupt the thin strata deposited by tides. Sedimentologists call this process of sediment disruption bioturbation. Organisms are so plentiful on some tidal flats that thousands of individual burrowers may be at work in only one square meter. In the Wadden Sea flats, burrowers destroy the bedding in all parts of the tidal flats except those areas where wave action or strong currents keep large amounts of sediment moving and prevent burrowing organisms from establishing themselves. Bioturbation is also very common in many relatively low-energy estuaries, such as those along the coasts of Georgia and South Carolina and most of the coast of the Gulf of Mexico.

Worm burrows are displayed above the sandy tidal flat in the Wadden Sea north of Wilhelmshaven, Germany. The shovel of sediment shows the vertical burrows.

ESTUARIES, SALT MARSHES, AND TIDAL FLATS

Where the tidal bedding and other physically formed structures are well preserved, burrowing organisms are absent. Worms and other soft-bodied creatures avoid the surface of tidal flats, where they are in danger of drying out. Filter feeders must avoid having their siphons clogged by abundant suspended sediment, so they cannot live where the water is very turbid. Burrowers cannot live in a mobile sediment bed either; when strong tidal action sweeps the sediment back and forth, the organisms are deprived of a reasonably stable substrate in which to burrow and maintain an existence. Clam diggers know that their best prospects on the flats are in the areas where tides are sluggish and waves are small or absent.

Salt Marshes

On the fringes of estuaries, lagoons, and other bays, in places where sediment deposits are sheltered from wave action and are above the level of neap high tide, vegetation eventually takes hold and forms a salt marsh. The upper limits of the salt marsh coincide with the landward or upper limit of the spring high tide, the highest level of regular inundation and sediment supply. A salt marsh may be a few meters wide or it may occupy the entire estuary except for the tidal channels.

An extensive marsh is a sign of a natural estuary that has largely filled with sediment. Many marshes mark the locations of old estuaries that have filled with sediment to the requisite elevation for plant germination. Some estuaries on the Georgia coast of the Atlantic have little open water; only tidal creeks dissect the extensive marsh environment. The Wadden Sea, by contrast, is too open to wave energy for vegetation to take hold except in the sheltered upper fringes along the mainland.

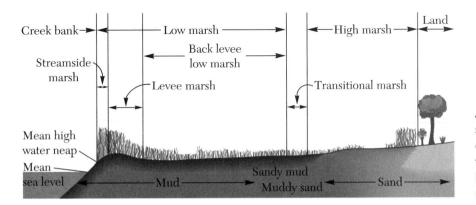

A cross-section of a salt marsh showing the subenvironments and distribution of marsh grass. From the edge of the tidal creek to the high marsh is cordgrass; the high marsh is needle rush.

Needle rush (Juncus) in the St. Mark's estuary in Florida.

In the upper intertidal region between the level of the neap high tide and that of the spring high tide—generally a short vertical distance of 20 to 50 cm—the velocity of the tide waters drops drastically and fine sediment settles without being disturbed by energetic waves. Opportunistic, salt-tolerant grasses, or in low latitudes, mangroves, are the first to take hold in this environment. The grass blades further slow tidal currents and trap more of the fine sediment carried in suspension. Once a stand of marsh grass is established, it becomes denser, thriving on its ability to tame the tides and accumulate sediment.

Their facility for trapping sediment, either by physically capturing it from passing currents or by slowing currents and permitting the sediment to settle into the plant community, makes these plants very important contributors to coastal sediment accumulation. In addition to their positive role in adding sediment to the substrate, marsh grasses are very important as sediment stabilizers. They prevent or inhibit currents and waves from removing sediment from the vegetated substrate.

The most common marsh grass in North America, Europe, and Africa is cordgrass, any of 16 species of the genus *Spartina*. A tough, long-leaved plant with spreading underground rhizomes that sprout new plants, it is an excellent sediment binder. Individual plants are generally about knee-high, but tall clumps at high elevations may be more than a meter. Salt marsh cordgrass is the most profusely growing plant in the world; tall stands along tidal creeks reach a mass of 30 tons per acre per year, making this one of the most productive of all natural environments and matching or exceeding the most intensely cultivated agricultural crops.

Cordgrass gives way to the needle rush, *Juncus,* and in some areas where water is fresh, the reed *Phragmites,* both of which have had numerous uses throughout recorded history—including furniture, roof thatching, musical instruments, and weapons such as blowpipes and arrows. The tall needle rush, *Juncus roemerianus,* has a tip sharp enough for hunting and dressing game. These reeds and rushes make up the high marsh, because they occur almost exclusively at the level of the spring high tide. The cordgrass, *Spartina,* characterizes the low marsh, the vegetated zone below the level of the spring high tide.

The development of an entire marsh can be characterized by the relative abundance of two common species of vegetation. Young marshes are dominated by cordgrass, with only a fringe of needle rush around the upper margin. Tidal channels are abundant and provide good avenues for tidal flux and sediment delivery. As a marsh matures, sufficient sediment collects in the upper intertidal zone to support more and more needle rush. In time, a marsh enters an intermediate stage of development in which the grasses and rushes are about equally distributed and tidal channels are fewer. In a mature marsh, sediment has become plentiful in the outer zones and only a few large tidal creeks interrupt a continuous stand of rushes. Eventually, land plants encroach on the outer edges of the rush stands. Although this is a paradigm for marsh development and the eventual filling of the estuary, the present global situation is one of eroding marshes due to sea level rise.

The marsh environment is quite similar to that of river and delta floodplains. Channels, bordered by natural levees and crevasse splays, cut through the marshy plain. Some channels meander and produce cutoffs and oxbow lakes. This system delivers sediment to the marsh in two ways: regular but slow flooding of the marsh by turbid water carried by sluggish currents that permit settling; and storm tides that push large amounts of

Oyster shells have accumulated, and a cordgrass marsh has begun developing, in the upper reaches of an estuary in Washington Oak State Gardens on the east coast of Florida.

sediment-laden water onto the marsh and deposit considerable sediment in a short time.

Eventually, an estuary fills with sediment. Its tidal flats give way to marsh, more and more of which remains above the mean high tide line; finally, terrestrial vegetation takes over from the grasses and rushes. For hundreds of years, the Dutch, Germans, and Danes have been hastening this process by converting marshes to farmland by draining them through a system of dams, dikes, and canals. Now the sediment has compacted and dried, and their reclaimed land is sinking—so they need higher dikes for protection. This process has now been stopped, and the Netherlands is trying to prevent further subsidence by flooding selected areas of farmland and returning them to marsh or open water for recreational boating.

Mangrove Swamps

In tropical and subtropical climates, extensive stands of mangroves—woody trees of various taxonomic groups—invade the intertidal zones of estuaries and other bays. Thick tangles of shrub and tree roots, commonly called swamps but properly known as mangles, form an almost impenetrable wall at about water level. Most trees grow from 2 to about 8 m high, some much higher, depending on the species and the environmental conditions—rare stands may be twice that high. The root system of mangroves are not only dense but also diverse in appearance and function. The most spectacular root display is put on by the red mangrove, *Rhizophora mangle,* which has large, reddish prop roots that support the tree. It also has vertical drop roots, which are long vertical appendages that sprout from low-lying branches and eventually reach the ground and give support. Pneumatophores, another common type of roots, occur in the other common species, the black mangrove, *Avicennia germinans.* The pneumatophore is a short root growing upward from lateral runners that

The massive root systems of mangroves create dense sediment stabilizing mazes.

extend from the central trunk. Although there is considerable argument about their function, it is believed that they are respiratory organs for exchange of oxygen. The third species of mangrove in Florida and the Caribbean is the white *Laguncularia racemosa.* In places like tropical Australia or India, there are more than 20 species of mangroves.

Mangroves cannot tolerate frost and are therefore limited in their latitudinal range. Although they only extend about halfway up the Florida peninsula on the Gulf Coast to about 28 degrees latitude, they are established south of Melbourne, Australia, along the Victorian coast up to 38.5 degrees latitude. These specially adapted trees can tolerate both fresh and salt water. In lowland estuaries, the mangrove trees may extend inland tens of kilometers along the estuarine river channel where the water is fresh.

A second factor limiting mangrove colonization is competition. Because they grow very slowly, mangroves are often outcompeted by marsh grass and other opportunistic plants.

The thickets of mangrove roots at the water line provide a sheltered habitat for a special community of organisms that are adapted to an environment intermediate between land and water. Barnacles and oysters encrust the roots and branches, looking almost like fruit. Fish, snails, and snakes all find protection, nesting sites, and food among the roots, which make a practically indestructible coastal defense against storms and hurricanes. The trees are destroyed by frost or by breakage during severe storms. Hurricane Andrew, in August 1992, sped across one of the largest mangrove areas in North America as it passed over southern Florida. Thousands of trees were broken about 2 m above the normal sea level and eventually died; some were also uprooted.

FOUR ESTUARIES

By way of summary, let us take a tour through four North American estuaries, each on a different kind of coast and each representative of a slightly different type. We begin at a low-energy coastal environment and proceed to a high-energy extreme that also has a large tidal range.

San Antonio Bay on the Gulf of Mexico

San Antonio Bay is on the central Texas coast about midway between Houston and Corpus Christi on the marginal sea coast of the Gulf of Mexico. It is shielded from the open Gulf by Matagorda Island, one of the many

Weak waves and tides have allowed a delta to form at the mouth of the Guadalupe River in San Antonio Bay, Texas coast.

Texas coast barrier islands. The Guadalupe River provides the single largest sediment load into the estuary. The estuary's gently sloping, soft bottom is only 3 m deep and is covered with a mixture of mud and sandy mud as well as widespread oyster reefs.

The tidal range at San Antonio Bay is less than 0.7 m. A fetch of at least 12 km in any direction allows modest wind-driven waves to provide most of the physical energy in the estuary and to cause complete mixing of the fresh- and saltwater masses. The wave energy and tidal influence are too weak to redistribute the sediment effectively, and a bayhead delta has developed at the mouth of the Guadalupe River. Infrequent hurricanes have influenced sediment distribution by winnowing fine sediments and concentrating the coarse shell debris into coarse storm layers that appear in the estuarine stratigraphy.

The bay supports a moderate but declining oyster industry. The extensive oyster reefs are typically oriented perpendicular to the tidal currents. This orientation allows maximum exposure to tidal flow, which provides the suspended detritus that nourishes these filter feeders. Their shells and fecal pellets are major contributors to estuarine sediments.

A narrow intertidal environment, a consequence of the low tidal range and a border of low eroded bluffs, supports a modest, discontinuous fringe of marsh containing mostly cordgrass (*Spartina*). Many other typical low-energy estuarine organisms besides oysters, especially worms and bivalves, are common and contribute to a thoroughly bioturbated substrate. This estuary, developed as sea level rose on a marginal coast, is really little more than a huge mud puddle.

Chesapeake Bay on the Central Atlantic Coast

Chesapeake Bay is situated on the stable trailing edge of the Atlantic coastal plain. Its elongate, complicated shape reflects its origin as a series of drowned river valleys along a rolling terrain of modest relief. The entrance to the estuary is between Cape Charles and Cape Henry, Virginia; this very high energy entrance has strong tidal currents and shifting sand shoals. The estuary extends northward for about 300 km to the mouth of the Susquehanna River. Other large- and modest-sized rivers empty into the estuary from the west—the Potomac, Rapahannock, York, and James—and smaller ones enter from the east. The several thousand kilometers of coast on the Chesapeake are both an asset and a curse, providing an extraordinarily rich and varied coastal area that is subject to extensive erosion.

The tidal range of Chesapeake Bay varies from about 2 m at the entrance to a barely perceptible change of water level in the northerly rivers. Wave energy varies greatly because of the widely varying fetches—from the entire length of the bay to that of the narrow arms of the river valleys. A strong north wind causes wave erosion around the city of Norfolk, Virginia, but much wave erosion also occurs at the protruding headlands scattered all over the estuary. Salinity gradients in this partially mixed estuary range from virtually zero at the river mouths to a nearly normal marine salinity of 35 parts per thousand at the bay entrance.

The estuary floor still bears the topography of the original drowned valleys, with considerable fine sediment infilling the lower areas. The nooks and small embayments of the eastern shorelines support extensive marshes. The western shorelines are straighter, often lined with low bluffs. Sediment carried by the rivers is relatively coarse and accumulates close to their mouths. The open part of the estuary is typically floored in mud, with a large infaunal population that is busy turning sediment over and producing pellets.

Chesapeake Bay has long been renowned for its thriving fish and shellfish populations; oysters have been especially abundant in the past. Since the 1970s, however, there has been a major deterioration in the quality of the bay environment. Suspended sediment and the estuary floor have become overloaded with sewage, toxic wastes, and heavy metals from the industries and cities along the rivers. Estuary life and commercial fishing have sharply declined; oyster harvesting is essentially finished.

Willapa Bay on the North Pacific Coast

Willapa Bay is situated on the Washington coast of the Pacific Ocean, along the tectonically active leading edge boundary between the North American plate and the Juan de Fuca plate. Not far inland is the Pacific Coast Range, a mountain chain with a steep westward drainage. Streams rush down the mountainsides directly to the narrow coast and terminate in a series of small estuaries.

Willapa Bay, the largest of these estuaries, is served by a deep and high-energy inlet. Much of the bay is protected from the open Pacific by a large barrier spit. The tidal range of nearly 4 m during the spring tides produces strong currents that carry a large sediment load. Mud accumulates in the deeper parts of the estuary basin and in protected areas around the margins. Tidal channels with large sandy bedforms are numerous and extend throughout much of the estuary. The extensive tidal flats expose about one-third of the bay area at low tide. Sand dominates near the mouth, which has changed its shape and position over historical time as wave and current regimes changed.

The large size of the estuary, nearly 50 km in its long direction, allows wind-driven waves to attain a modest size and, along with large tidal flux, to mix the fresh and salt waters of the estuary. Salinities are high for an estuary, in the range of 20 to 30 parts per thousand, because the volume of seawater swept in with the tides overwhelms the modest freshwater input. Nevertheless, oysters thrive in this estuary and support a vibrant industry, which markets both meat and shells.

Much of Willapa Bay is bordered by cliffs. The waters of high tide reach their bases or a bit higher. Marshes therefore are restricted to those small coves where the tides have deposited sediment and to the estuary side of the barrier spit across the mouth. This is one of the most beautiful and relatively undiscovered estuarine coastal areas of North America.

Bay of Fundy on the North Atlantic Coast

The Bay of Fundy, on the southeastern Canadian coast, cuts inland for 150 km between the provinces of New Brunswick and Nova Scotia. This funnel-shaped estuary is the most renowned in the world because of its spectacular tides, the highest in the world. As the tidal range increases up the estuary, there is total mixing. The huge volume of seawater entering at the mouth completely snuffs the effects of the modest freshwater influx. Salinities are high—20 to 32 parts per thousand is typical, even in the upper Minas Basin and Cobequid Bay.

The Bay of Fundy has other distinctions as well. It is set in a basin that is bounded on each side by faults that appeared at the time North America separated from Eurasia during the Triassic breakup of the supercontinent Pangaea. The steep cliffs that line the estuary include 200-million-year-old strata called red beds because of their bright color. Laid down in a hot, dry climate, the red beds are now mixed and overlain with glacial debris of the Pleistocene ice age, much of which is derived from the older red beds and also carries a strong red color. The modern sediments accumulating in the bay give the entire area—water, sediments, and rocks—an unusual color. When the tide is running out, it is streaked with suspended red mud from the erosion of the bedrock and the glacial sediments.

Numerous tree stumps along the intertidal flats of the bay testify that the area was once terrestrial upland. Radiometric dating of the stumps and other terrestrial organic material has established that the tidal range of the bay has been increasing by up to 30 cm (1 ft.) each century for the past 6000 years and has more than doubled in only 3000 years. The dramatic change is due in part to rebound, an isostatic uplift of the continental plate as the thick overlying glacier melted. The rebound has caused the floor of the upper end of the bay to rise and thereby force the same amount of water into what is essentially a smaller container. The low tide position has remained nearly the same, but because of the effect of the rebound, the high tide level has increased greatly.

The floor of the estuary is composed mainly of coarse red sediment and gravel mixed with sand and mud. The strong tidal currents that rush up the sloping bay form a distinct pattern of decreasing grain size toward the landward end of the estuary near the river mouths. Soft thick mud dominates the river mouths and forms a narrow band around the shore near the high tide line. This profile represents a rather condensed version of the typical grain size trend that results from decreasing energy of tidal currents and causes grain size to decrease up the slope of the intertidal zone. The large tidal range and the steep bay margin produce this condensed trend, which generally culminates in a narrow, nearly discontinuous strip of *Spartina* marsh at the top. The lack of continuous marsh growth is due to the absence of flat land combined with the tide-driven mobility of the sediment substrate.

The stump of a large tree on a tidal flat in the Bay of Fundy is evidence of the recent increase in tidal range in this famous estuary.

The roiling of the sediment during tidal movements creates a difficult environment for benthic organisms. The intertidal region of the Bay of Fundy is peculiarly devoid of burrowing creatures and shellfish. The few inhabitants are mobile creatures that can scurry away from the tides and withstand the conditions of a highly mobile substrate. As one walks over the sandy flats for long distances, one sees little disturbance of the surface. Burrows that are typically so common on tidal flats are missing, and few gulls or shorebirds fly overhead, for there is no food for them to feast on here.

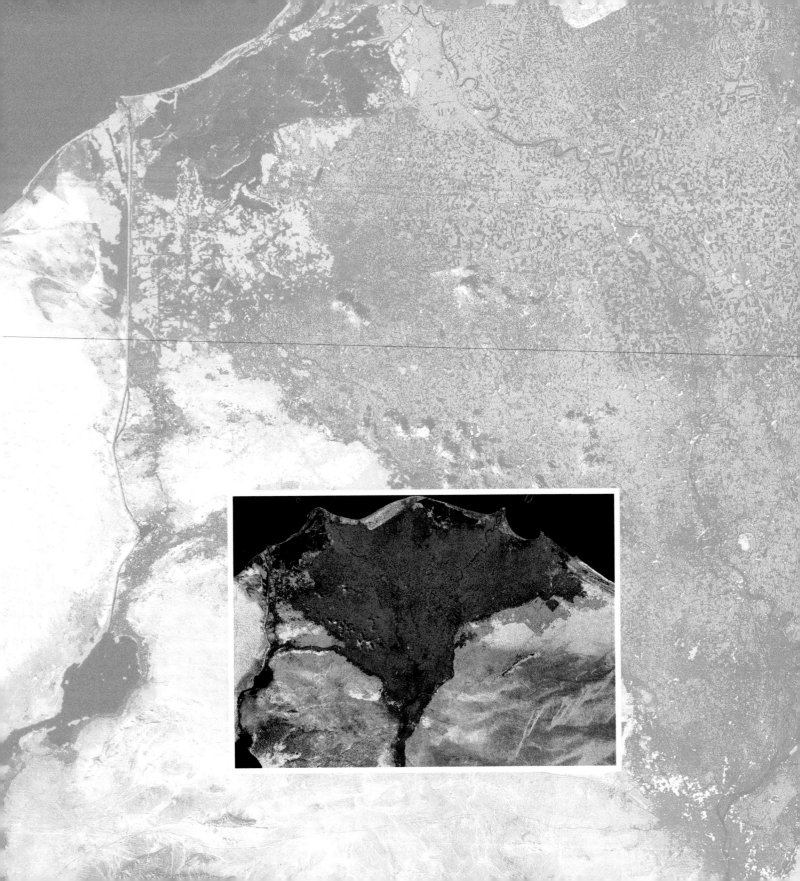

5

Deltas—Where Rivers Unload Their Deposits

A delta begins at the "point" where a large, sediment-laden river leaves its upland drainage basin and flows onto a more level region adjacent to the ocean. On this plain, the river slows and usually divides. Most rivers break up into channels called distributaries because the water and sediment flow outward, or are distributed, from the main channel. As the distributaries continue to branch, the delta region takes on the shape of a triangle, with the undivided river at its apex and the shoreline at its base. About 450 B.C., the Greek historian Herodotus had the insight to recognize that the vast area bounded by the Nile and its distributaries was shaped like the Greek letter delta (Δ), and he gave this landform its aptly descriptive name.

Deltas are transitional coastal environments that are neither fully terrestrial nor fully marine. They have no easily recognizable landward or seaward boundaries but change by imperceptible stages from open sea to solid ground. Built primarily from riverborne sediment, deltas form when the amount of sediment delivered at the mouth of a river exceeds the amount removed by waves and tidal currents. Deltas develop best where rivers deposit sediment loads onto low-energy continental margin.

A satellite image of the Nile delta vividly displays the vegetation, here colored red, on the rich soils of the delta.

In their mutability and transitional nature, deltas resemble estuarine bays. Both estuaries and river deltas are strongly influenced by rivers, waves, and tides; both are important sites of sediment accumulation; both have tidal flats, marshes, and swamps; and both are geologically young features that are vulnerable to eustatic and local changes in sea level. In fact, either term may be correctly applied to the same location, for some rivers build deltas within their estuaries.

In terms of coastal morphology, however, deltas and estuaries are opposites. Whereas an estuary is an arm of the ocean extending into a river, a delta is an arm of the land projecting into the receiving basin, building outward into the sea as great quantities of sediment are delivered by the river.

WHERE DO DELTAS DEVELOP?

Deltas occur on every continent and in a wide range of climatic settings, but the geologic settings are generally similar. A tectonically stable trailing edge coast provides the right conditions for delta formation. It has low- to moderate-relief terrains, such as coastal plains or geologically old mountain areas. Rivers bring an abundant sediment supply across wide, gently sloping land, where the river channels meander back and forth on their way to the coast. On the seaward side of this tectonic setting, the broad continental shelf provides a platform suitable for sediment accumulation; it also reduces the size and energy of the incoming waves. The São Francisco delta in South America and the Senegal delta in Africa have developed on trailing edge coasts.

Not all trailing edge coasts develop major deltas, however. The Atlantic coast of the United States has many substantial rivers, but it has no significant deltas. Nearly all of the east coast rivers empty into estuaries such as the Chesapeake Bay and the Delaware Bay, where sediment is generally dispersed by tidal currents.

Marginal seas with trailing-edge characteristics provide shelter from large waves and tides, and very large deltas have developed in tectonic settings of this type. Excellent examples are the Mississippi River delta in the Gulf of Mexico; the Rhone, Nile, Po, and Ebro deltas in the Mediterranean Sea; and the huge deltas of China that empty into the South China Sea.

There are no significant deltas on the western margins of North and South America, where the tectonic setting is not conducive to delta formation. These coasts are converging margins where an oceanic plate is being

Large Modern Deltas

River	Landmass	Receiving Basin	Size (km²)	Annual Sediment Discharge (tons × 10⁶)
Amazon	South America	Atlantic	467,000	1,200
Chao Phraya	Asia	Gulf of Siam	11,000	5
Danube	Europe	Black Sea	2,700	67
Ebro	Europe	Mediterranean	600	?
Ganges–Brahmaputra	Asia	Bay of Bengal	106,000	1,670
Huang	Asia	Yellow Sea	36,000	1,080
Irrawaddy	Asia	Bay of Bengal	21,000	285
Mahakam	Borneo	Makassar Strait	5,000	8
Mekong	Asia	South China Sea	94,000	160
Mississippi	North America	Gulf of Mexico	29,000	210(469)
Niger	Africa	Gulf of Guinea	19,000	40
Nile	Africa	Mediterranean	12,500	0(54)
Orinoco	South America	Atlantic	20,600	210
Po	Europe	Adriatic Sea	13,400	61
Rio Grande	North America	Gulf of Mexico	8,000	17
São Francisco	South America	Atlantic	700	6
Senegal	Africa	Atlantic	4,300	?
Yangtze	Asia	East China Sea	66,700	478

Data in parentheses are discharge volumes prior to major influence of dams and other human influence. Data from Wright, 1982, and Milliman and Mead, 1983. Wright, L.D., 1982, "Deltas," in Schwartz, M.L. ed., Encyclopedia of Beaches and Coastal Environments, Huthinson Ross Publ. Co., Stroudsburg, Penn. (p. 359) p. 358–368. Milliman, J.D. and Meade, R.H., 1983, World-wide delivery of river sediment to the oceans, J. Geology, v. 91, p. 1–21.

subducted under a continental plate. Such coasts abut mountainous and high-relief terrain, where rivers are usually narrow and drain through steep valleys. The high-relief drainage system along convergent coasts provides a limited area for eroded sediment to be collected for delivery to the coast. Although the steep valleys erode rapidly and their rate of sediment production is high, the total amount provided to the river mouth is small relative to that found on trailing-edge coasts. Additionally, when the rivers reach the coast, the coastal cliffs and steep continental shelf provide an insufficient platform for sediment accumulation. Instead, crashing waves scour the coast of most of the sediment load that arrives. The Columbia River, among the largest streams to empty into the eastern Pacific, originates in a produc-

The influence of waves has formed the winglike spits on either side of the Ebro delta, on the Spanish coast of the Mediterranean Sea, seen here in a satellite image.

tive upland area and carries a substantial load of sediment to the coast, but strong waves and currents prevent delta formation. Elsewhere, large submarine canyons near river mouths channel sediment from the surf zone directly to deeper parts of the ocean.

How Old Are Our Deltas?

Most of the present active deltas are very young geologic features; some are only a few hundred years old. Because a delta develops at the coast, its existence is, in part, controlled by sea level. During the periods of extensive glaciation, sea levels were much lower and rivers traversed the present continental shelves, dumping their sediment loads at or near the outer shelf edges. This suspended sediment cascaded down the continental slopes in turbulent, high density flows called turbidity currents. New deltas did not

DELTAS—WHERE RIVERS UNLOAD THEIR DEPOSITS

form during this period, and deltas that had previously existed near the positions of present-day coasts were abandoned and entrenched by rivers as they flowed across the continental shelves.

Melting glaciers brought a rapid rise of sea level, and river mouths retreated so rapidly that deltas could not develop. Finally, about 7000 years ago, the Holocene sea level rise slowed, and in some parts of the world it stabilized at approximately its present position. Where conditions were appropriate, deltas began to develop as large quantities of river sediment accumulated.

Not all present-day deltas are only a few thousand years old. Many of them have formed on ancestral deltas built up during previous interglacial periods. A few, such as the Mississippi and Niger deltas, are underlain by ancestral deltas that formed tens of millions of years ago. The upper regions of these mature deltas are also ancient, but their active delta lobes are only between 3000 and 6000 years old. The lower Mississippi delta includes

Each time the Mississippi River shifts its channel, the river delta develops a new lobe. Over the past 7000 years seven major lobes developed, numbered here in chronological order, and several additional lesser lobes.

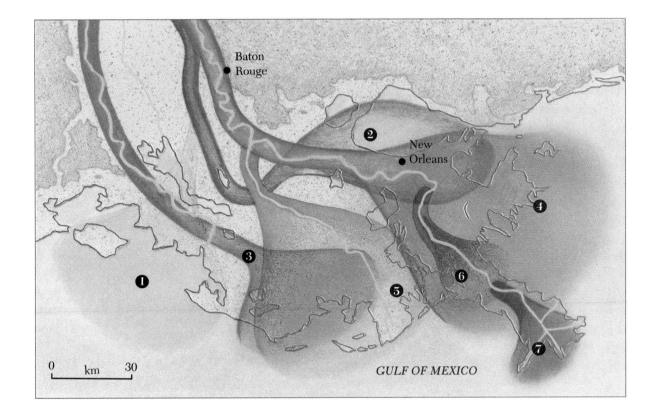

16 detectable lobes. A new lobe forms whenever the location of the river mouth changes. The channels of abandoned lobes fill up with sediment contributed both by the river and by the waves and tides of the coast. The present delta lobe of the Mississippi dates back only 600 years; its most active portion has developed since New Orleans was founded in 1717.

ENVIRONMENTS OF THE DELTA

The formation and extent of specific environments within a delta depend on the interaction between the flow and distribution of the river's sediment and the wave and tidal currents of the ocean's margin. As the water flows from the river's mouth, its velocity decreases and it loses its capacity to carry sediment. Consequently, sediments accumulate in the river-mouth area and the river divides into distributary channels. Each distributary channel then continues to transfer massive amounts of fine-grained sediment to the coastal area.

The river carrying the greatest sediment load is the Huang (Yellow River), originating in fast-flowing streams of the loam-rich highlands in north-central China. Its waters contain nearly 80 g of sediment per liter (about 2 lbs per cubit foot); flood discharge is sometimes 70 percent sediment. When this turbulent, muddy river reaches the coastal plain, its chan-

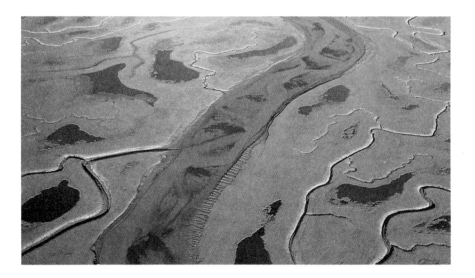

Marshes and bays fill the areas between distributaries on either side of a sediment-choked channel running through the Copper River, Alaska delta plain.

DELTAS—WHERE RIVERS UNLOAD THEIR DEPOSITS

nels fan out into a delta and its sediment-laden waters sustain vast, partially submerged rice fields.

The landward and very flat part of a delta is the delta plain. The upper delta plain is merely an extension of the upland meandering river system, except that the river here consists of one or more distributary channels.

Each time a distributary channel overflows its banks, the coarser sandy sediment particles are dumped first, producing a low ridge of accumulated sediment along the bank margin. This ridge is the natural levee. It may build up to an elevation of a meter or two above the surrounding delta plain. During subsequent flooding, the natural levee may be breached either through a naturally low section or through cuts made for human passage. When the sediment-laden floodwaters pass through the breach, generally called a crevasse, there is an immediate reduction in carrying capacity as their velocity decreases abruptly. A thin, fan-shaped sediment accumulation forms beyond the breach. This formation, called a crevasse splay, can extend several kilometers across the upper delta plain.

As the distributary channels approach the ocean, the gradient of the land flattens even more and the height of the natural levees decreases. The delta changes into a domain of marsh grass, shifting channels, and brackish interdistributary bays. In warm climates, mangrove swamps flourish in the same niche generally occupied by salt marshes. In all parts of the world, the delta system teems with plant and animal life; it is among the most productive of environments. Not only are deltas rich in shellfish, market fish,

Bands of vegetation outline numerous natural levees on the flooded outer Mississippi delta. In the foreground is a large crevasse splay where a levee has been breached.

water birds, and other animals, they are also a primary nursery ground for animals that later migrate to the open ocean.

When floodwaters rise high enough to spill over the levees and beyond the splays, they cover the interdistributary area with muddy water. On the upper delta plain, this sediment contributes to floodplain deposition, but in the lower delta plain, the mud settles into the interdistributary bays and marshes. Eventually, this process fills all the interdistributary bays with sediment and the silt-filled bays become vegetation-covered marshes.

The inside edges of the bends in the distributary channels on a delta plain fill with thick accumulations of sand and gravel. These deposits are called point bars. As the channels migrate across the delta plain, they leave subtle but recognizable scars marking their former locations.

The major landforms of the delta plain—natural levee, crevasse splay, interdistributary bay and marsh—are distinguished from one another on the basis of elevation, sediment character, and vegetation. As time passes, continued flooding and sediment unloading enlarge the delta and bring

A false color satellite image of the modern lobe of the Mississippi delta showing the bird's-foot configuration with sediment plumes discharging from the distributary mouths.

DELTAS—WHERE RIVERS UNLOAD THEIR DEPOSITS

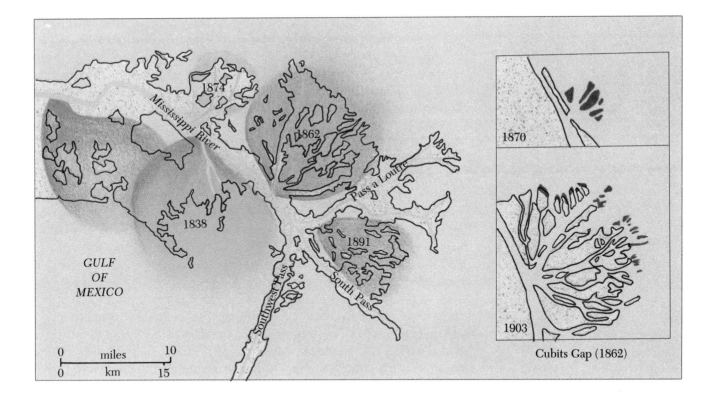

Cubits Gap (1862)

more and more of its features above water level. Much of the mature delta plain between the distributary channels eventually turns into fertile farmland interspersed with small lakes and freshwater marshes and swamps. All these are periodically replenished by floodwaters.

The seaward edge of the delta plain merges with the subtidal portion of the delta—the delta front—where wave and tidal energies affect coastal development. Here, sediment empties out of the mouth of each distributary channel as both suspended load and bed load. The muddy (fine) suspended sediment tends to be carried away from the mouth of the channel by wave and tidal currents, whereas much of the sandy bed sediment tends to accumulate near the channel mouth.

The amount and configuration of sand accumulations in the delta front depend on the relative roles of the interacting river, wave, and tidal currents. A common type of sand accumulation is the distributary mouth bar, a sandbar that accumulates just seaward of the channel mouth and typically causes the channel to bifurcate. Waves that approach the delta at an oblique angle create currents that carry this sand from the mouth of the

Growth of a delta along a distributary channel through the sequential building of crevasse splay deposits.

channel and distribute it along the outer delta plain, thereby forming a nearly continuous system of sandbars.

At the seaward edge of the delta front, the suspended sediment in the river water finally settles out into the deeper coastal water as the velocity of the outflowing river water slows. This mud accumulation is generally very thick and extends across part of the continental shelf.

The Contest Between the River and the Sea

At the front edge of the delta, the fate of the sediment and of the delta itself is decided over time by the relative influences of the river from the landward side of the delta and the marine processes on the seaward side. Strong tidal currents that wash over the interdistributary landforms may scour them of sediment, carrying the load oceanward and piling it up in long sand bodies perpendicular to the shore. Energetic waves may cause strong longshore currents to sweep sediments along a rather smooth shoreline. When the river flow is dominant, it overcomes these marine processes and pours enough sediment seaward to prograde the delta further into the ocean. When marine processes dominate over the river, however, sediment is carried away and the delta erodes.

For a delta to grow or at least maintain its size, the river must be sufficiently vigorous and carry enough sediment to keep the marine processes in check. But the amounts of water and sediment a river discharges depend on climate, topography, and sediment availability in the drainage basin, the gradient of the river, and a host of local variables. Nevertheless, the most important factors for the vitality of a delta are rainfall and soil type. Prolonged drought in a river's drainage basin is devastating to its delta. The Ord River delta in northwestern Australia is presently succumbing to waves and tide for this reason. By contrast, profuse rainfall and loose soils are a delta's salvation. The large and growing deltas of today all experience periodic heavy floods that ensure their survival—but are ruinous for the people who live on them. The Huang and Ganges–Brahmaputra deltas are still growing for this reason.

Human activity also has a tremendous influence on delta conditions because of our ability to control river water and sediment discharge. Loss of sediment upstream of the river mouth leads to delta erosion. The Nile River delta is experiencing serious problems because damming of the river has led to a cessation of sediment delivery.

DELTA SHAPES

The combined effects of waves, tides, and river on the distribution of sediment determine the overall shape of a delta. Each delta reflects the struggle between the rate of sediment delivery and the form of its dispersal, and the three forces that compete to give the delta its distinctive shape. In 1975 William Galloway, a geology professor at the University of Texas at Austin, saw that the characteristic shapes of deltas could be the basis for a broad classification scheme reflecting the interaction of these forces. His system,

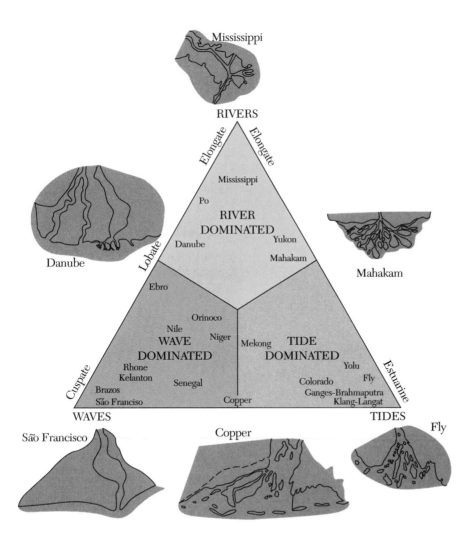

William Galloway's triangular diagram classifies river deltas according to the relative influence of the three major factors affecting their development: the river, waves, and tides.

which has since become the standard, is presented as a triangle, with the apex representing river (fluvial) influence and the corners of the base the wave and tidal influences. The size of the delta and the absolute contribution of each force to its shape are unimportant in determining its position within the triangle. What counts are the relative influences of the three forces interacting over time. Any delta that is strongly affected by only one of the three factors is placed in the appropriate corner. A delta with a balanced interaction of forces is placed near the middle of the triangular diagram.

River-dominated Deltas

In general, river-dominated deltas have a well-developed delta plain with several distributaries projecting seaward in a digitate, "bird's-foot" configuration. Conditions that foster river-dominated deltas include an ample flow of freshwater and sediment from the upland and a relatively placid seaward receiving basin. A marginal sea that has small waves and a small tidal range and is adjacent to a tectonically stable coast best meets these conditions.

The Gulf of Mexico offers a perfect setting for a river-dominated delta, and the Mississippi River delta has a perfect bird's-foot shape. Other outstanding river-dominated deltas are the Danube on the Black Sea, the Po on the Adriatic, and the Huang (Yellow River) on the Yellow Sea. All have good sediment supplies from the drainage basin, a small range between high tide and low tide, and a receiving basin that is sheltered from large waves. The Mississippi, however, is in the path of hurricanes, and over the years devastating storm surges have caused accelerated erosion on the delta front. Such storms do not, however, control the overall delta configuration.

Tide-dominated Deltas

Tide-dominated deltas develop where a large range between the high and low tides leads to strong tidal currents that flow essentially perpendicular to the coast. The wave height is moderate to low on these deltas, and, as a result, longshore currents are weak. Consequently, the freshwater discharge is overpowered by tidal currents intruding from the seaward direction. These strong tidal currents mold the sandy delta sediments into elongate landforms that are generally parallel to the river flow and perpendicular to the general trend of the coast. Marine sediments carried inland by the tidal currents may outweigh river sediment, and salt marshes

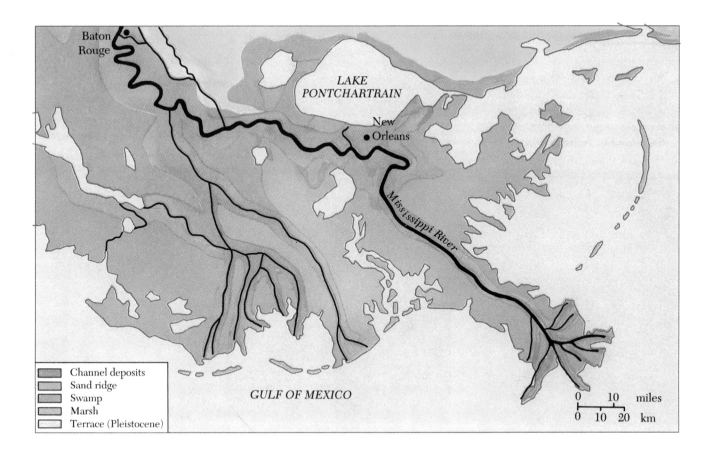

Legend:
- Channel deposits
- Sand ridge
- Swamp
- Marsh
- Terrace (Pleistocene)

Baton Rouge

LAKE PONTCHARTRAIN

New Orleans

Mississippi River

GULF OF MEXICO

0   10   miles
0   10   20   km

rim the typically extensive intertidal flats. Tide-dominated deltas resemble estuaries because of their embayed setting of salt marshes, swamps, and tidal flats.

The largest of the tide-dominated deltas is the Ganges–Brahmaputra delta on the Bay of Bengal. Fed by two large river systems, the delta channels carry huge amounts of sediment to a coast characterized by modest waves, a tidal range exceeding 3 m, and monsoonal floods. The seasonal load brought down to the coast during the wet monsoon season is ten times greater than that supplied during the dry season. Strong tidal currents flow inland in numerous large channels, where the sediment is redistributed into long bars perpendicular to the coast.

The Mississippi River delta is a good example of a river-dominated delta.

Dominating tidal currents have formed an estuary-like inlet at the mouth of the Ganges River, seen in this satellite image of a portion of the Ganges–Brahmaputra delta near Dacca, Bangladesh. The tips of elongate bars running parallel to the flow can be seen in the lower right corner.

Wave-dominated Deltas

Wave-dominated deltas typically have a rather smooth shoreline with well-developed beaches and dunes. The delta plain tends to have few distributaries; some deltas of this type have only a single channel. As the channel delivers its sediment load to the basin, longshore currents carry sandy sediment along the smooth outer shore and the finer sediment is transported even further away. A wave-dominated delta is generally smaller than other types because the distributing power of the waves striking the delta front is stronger than the carrying power of the river. Were the wave processes to become strong enough to carry all the river sediment away, the delta would shrink and eventually disappear.

Two different shapes characterize wave-dominated deltas. Which shape develops depends on the way the longshore currents sculpt the river sediment. The São Francisco delta in southern Brazil is built up on either side of its single channel, which empties into the Atlantic Ocean. The general shape of the delta is symmetrically cuspate because there is no strong littoral drift in either direction. The Senegal River on the eastern coast of the Atlantic in Africa typifies the other wave-dominated shape. There the predominant winds produce waves that approach the coast from the north and generate a strong southerly longshore current. The resulting littoral drift has deflected the river course over 50 km and created a long sand spit, the Barbary Tongue. This barrier spit protects the extensive wetlands that cover the delta plain.

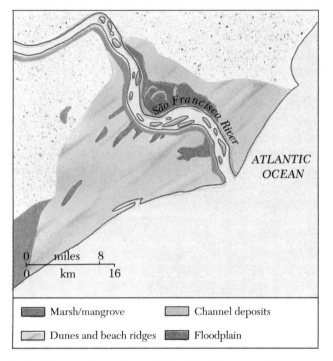

Marsh/mangrove   Channel deposits

Dunes and beach ridges   Floodplain

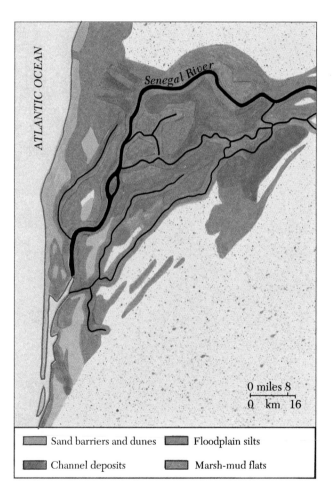

Sand barriers and dunes   Floodplain silts

Channel deposits   Marsh-mud flats

Left: The moderate flow of the São Francisco River, with its single channel, and the weak tides of the southern coast of Brazil have allowed the waves to shape the São Francisco delta. The result is the delta's cuspate shape with well-developed beaches and dunes on the outer delta.

Right: Another wave-dominated delta is the Senegal on the coast of west Africa. There, the strong north to south littoral drift has forced the main channel to migrate more than 50 km to the south.

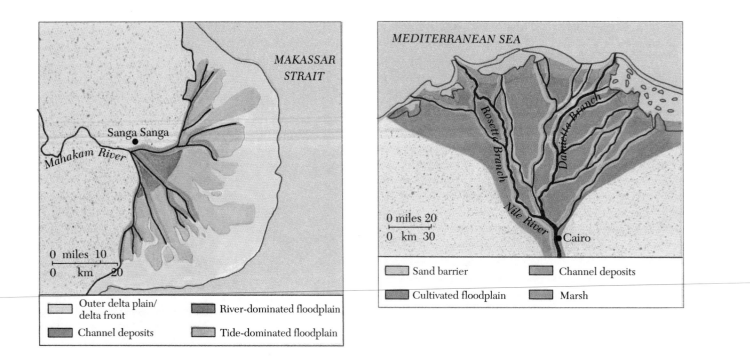

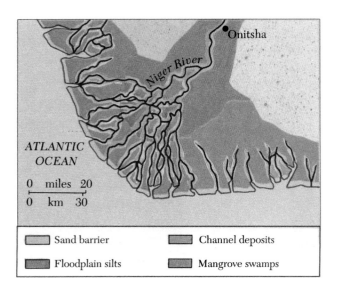

Top left: Tidal currents have scoured the bird's-foot Mahakam delta on the Pacific Ocean, both at the distributary mouths and in between, creating numerous marshy lobes that lack distributary channels.

Top right: Waves have created the even and sharp outer boundary of the Nile delta by smoothing the coast between distributary mouths.

Left: The Niger delta on the west coast of Africa represents a good mixture of river, wave, and tidal influence.

Intermediate Delta Types

Placed somewhere toward the middle of Galloway's schematic triangle are deltas shaped by rivers, waves, and tides in various combinations. In a river- and tide-influenced delta, the distributary channels project seaward in bird's-foot lobes while the tidal currents scour their way between the "toes." The Mahakam delta on the Borneo coast of Makassar Strait is a good example. The powerful Mahakam River, with numerous distributaries, is influenced by currents produced by tides with a 3-m vertical range across a wide delta front. Wave action in the strait is low.

The Nile delta shows a river- and wave-dominated front. A small number of well-defined distributaries traverse the delta plain and project into the Mediterranean Sea. The tidal range is small, and waves are sufficiently energetic to smooth the coast between the distributary mouths into discontinuous calm beaches and low dunes separated by muddy marshes and tidal flats.

Near the center of the classification of deltas, with almost equal influences of river, waves, and tides, is the Niger delta on the Gulf of Guinea in western Africa. The delta plain is well developed with a complex network of distributaries—features of river-influenced deltas. The broad lower plains with numerous tidal channels reflect the 2.8-m tidal range. High-energy offshore waves have smoothed the delta front and created storm-beach ridges, giving this delta the cuspate character of a wave-influenced delta.

HUMAN INTERVENTION

Deltas have always been choice places of human habitation despite periodic flooding. Early civilizations developed on deltas to take advantage of both the fertile soil and the nearby fisheries. Access to the sea also provided military advantages as well as avenues for trading.

As delta regions became more populated, upland forest clearing and cultivation usually advanced delta growth by intensifying soil erosion—the extra sediment carried away in streams and rivers expanded many small deltas and some larger ones. We see an extreme example of this development today in the Amazon delta. Until recent years, the river, despite its great size and sediment load, was only able to establish a modest delta on the Atlantic against the strong waves and high tides of the coast. Now the delta is growing rapidly, nourished by vast quantities of new sediment washed out of cleared rainforest regions.

A brown plume of sediment flows into the ocean near the mouth of the Amazon River in Brazil. The plume is evidence of the tremendous amount of suspended sediment being washed out of the rainforest.

The more common effect of human activity, however, is to shrink the size of a delta by diminishing the velocity of its river and depriving it of sediment. River discharge is reduced by diverting water for crop irrigation and city water supplies. It is also reduced by navigational locks and by hydroelectric dams and reservoirs; these structures also act as important sediment traps.

The Aswân High Dam in Egypt illustrates the dilemma encountered in trying to control the discharge of a major river system. Completed in 1970, the dam has ended destructive flooding by controlling the release of the

Flood conditions in the New Orleans suburb of Westwego in the 1890s. Now the river discharge in the New Orleans area is controlled and much of the flood discharge is released upstream into Atchafalaya Bay.

DELTAS—WHERE RIVERS UNLOAD THEIR DEPOSITS

annual Nile floodwaters. It irrigates millions of acres of former desert, generates a large amount of electric power, provides improved navigation, and supports a fishing industry in its 22-km-long reservoir. Nevertheless, the fertility of the delta land is declining because silt is trapped in the reservoir and is not being made available for distribution to the delta downstream, where agriculture is extensive. The delta front itself, deprived of sediment, is being eroded by waves from the eastern Mediterranean Sea. The fishing grounds in the eastern Mediterranean are in decline because of the absence of nutrient-laden Nile River water and because of the destruction of delta marshland, which is an important spawning and nursery ground.

A similar condition has taken place at the Colorado River delta located at the north end of the Sea of California separating Baja California from the rest of Mexico. The headwaters of the Colorado are in the Rocky Mountains in the state from which it is named. Along the course of over a thousand kilometers, there are numerous dams and reservoirs and sites where water is diverted for human use, primarily to southern California. Consequently, virtually no water or sediment is currently being provided to the delta, which is rapidly being eroded by strong tidal currents.

Once a river and its delta have been tamed by giant engineering projects, it seems almost impossible to return them to a more natural state; and all attempts to control deltas have weakened them. The questions now being asked are, What are deltas worth? And what is best for the huge populations who depend on them?

Looking south, toward the sea, over a portion of the Colorado River delta on the coast of the Sea of California, at low tide. The extensive, vegetation-free tidal flats are the result of desert conditions and a high tidal range.

6

Beaches, Dunes, and Barriers

When we think of the coast, the beach is often the first coastal element that comes to mind, closely followed by the dunes that are landward of the beach. Both beaches and dunes are found on the mainland and on barriers, although their features are different in these two locations.

The beach acts as the seaward protection for the coast in general, whether it is part of a barrier island or on the mainland. It is the most actively changing part of the coast—each wave shifts its sediment.

Barrier islands and other barriers characterize much of the coast along trailing edge margins, as well as on scattered parts of leading edge margins. They form as sand accumulates through the combined action of waves and wave-generated longshore currents. Barrier coasts include, but are not limited to, barrier beaches, barrier spits, and barrier islands. These narrow, elongate accumulations of sediment rise above sea level, offering the landward part of the coast a natural source of protection from wave attack.

Ridge and runnel system along a barrier island in the southeast United States coast showing the ridge that is migrating shoreward over the rippled runnel.

BEACHES

The nearshore—that is, the shallow marine water adjacent to the beach—is the portion of the coast that attracts surfers, sportsfishers, and those who enjoy the beauty of the breaking waves. It extends seaward from the lowest tide line at the seaward extent of the beach, out across the surf zone, which typically includes sandbars. Although some nearshore areas are smooth and gently sloping, most comprise a combination of sandbars and intervening troughs on the seaward side of the surf zone. In many places, a pair of persistent sandbars, over which waves break during storms, parallels the beach.

The beach extends from the low tide line landward across the unvegetated sediment to the beginning of permanent vegetation or to the next geomorphic feature in the landward direction—a naturally occurring dune, a rocky cliff, or a constructed seawall. Sandy beaches include a foreshore, a backshore, and, sometimes, a storm ridge. The foreshore includes the intertidal zone and extends to the landward break in the slope, the berm. It typically slopes gently toward the sea and may display a small ephemeral bar and trough morphology, variously called a ridge and runnel or a swash bar. The foreshore includes the swash zone, where waves rush up and back across the shoreline.

The backshore, or backbeach, extends from the berm at the landward end of the foreshore across the remainder of the beach. This portion of the beach is nearly horizontal or slopes down slightly in the landward direction. It is open to the wind, but waves reach it only during storm surges. After major periods of erosion, the backbeach may be greatly reduced or even absent.

Gravel beaches of shell and rock fragments commonly include a storm ridge that is just landward of the foreshore. This feature may grow until it rises several meters above high tide and entirely replaces the backbeach. Its composition depends on the nature of the gravel material in the immediate area; its size is proportional to the rigor of the storms that produce it.

The overall profile of a beach and the adjacent nearshore depends on sediment supply, wave climate, overall slope of the inner continental shelf, perhaps tidal range, and a variety of subtle local conditions. Reflective beaches are steep beaches with poorly developed nearshore sandbars or none at all. Because of their steepness, beaches of this type reflect most of the energy of the waves that strike them. And, because the nearshore slope is also steep and free of sandbars, the waves arrive at the beach with full strength. Both these factors contribute to the erosive conditions that prevent the development of sandbars. At the other extreme is the dissipative

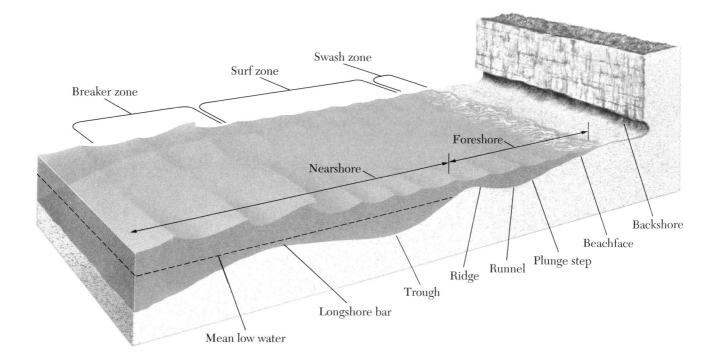

beach, a gently sloping nearshore and beach where wave energy is gradually lost as the waves move toward the shore. The gentle wave action, however, may also promote sandbar development. Well-developed nearshore bars and rip channels are most commonly associated with beaches that have moderate slopes.

Beach Materials

Nearly anything that can be transported by waves can form a beach. The available materials will be incorporated as long as they can be carried to the beach and remain there, at least temporarily. For example, mud is an uncommon primary component of a beach, but where it is the dominant sediment available and where wave climate permits it to be stable, it can accumulate in large quantities. The most extensive and best known mud beaches are along the coast of Surinam, on the north coast of South America, where huge quantities of fine sediment are imported from the Amazon River. The low wave climate and abundant mud produce beaches with the consistency of yogurt.

The beach extends seaward from a cliff or other geomorphic feature, through the backshore, to the end of the intertidal zone at the landward edge of the nearshore.

Pebbles from glacial deposits and eroded bedrock make up a beach at Olympic National Park, Washington.

Given the correct set of circumstances, gravel particles of virtually any size and composition can be a common beach component. All that is needed is an easily available source of rock fragments and a high wave climate. Gravel beaches range from small pocket beaches between headlands on an otherwise rocky coast to broad stony beaches that extend for kilometers—in the northern latitudes the latter are usually pebbles and cobbles from glacial deposits and stream accumulations or rock fragments directly eroded from bedrock. Many gravel beaches associated with barrier islands are composed of gravel-sized shells. These are especially abundant in low to mid-latitudes and in other places where there is little other sediment available.

The sand of most beaches is composed mainly of quartz, but many other minerals are present in varying quantities. A few beaches have sand composed of only one type of mineral or rock fragment, because that is the only material available. Examples include the black or green volcanic beaches of some Pacific islands and the fine white shell beaches of the Bahamas and southern Australia.

As a consequence of wave activity, the range of particle sizes found on a given beach or portion of the beach is narrow; that is, the waves and their related processes arrange the sediment particles in such a fashion that at any one place the sediment particles are more or less the same size. This phenomenon is called good sorting. The same observation is true for gravel beaches; they are well sorted even though the absolute particle size (greater than 2 mm) is much larger than that of sand beaches ($\frac{1}{16}$ to 2 mm).

Beach Processes

Waves, tides, and the currents they generate influence both the sediment and the structure of beaches. The role of tides and tidal currents on beaches is subtle. As the water level rises and falls, the shoreline moves. This in turn continuously changes the portion of the beach influenced by the waves; and the greater the tidal range, the wider the band of change. Tidal currents, although important agents in other settings, are so weak in the shoreline area that they appear to have no significant effect on beaches.

Waves breaking against the shore obviously interact with beach sediment. Each crash of the wave temporarily suspends an amount of sediment directly proportional to the size of the wave. This suspended sediment is then moved by currents—primarily wave-generated, longshore currents.

Waves produce four types of currents: combined flow currents, longshore currents, rip currents, and swash zone currents. All four occur in the surf zone and adjacent beach, and all move sediment.

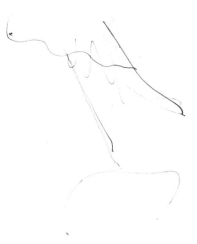

The coarse gravel on this beach at Big Sur, California, is well sorted and well rounded, like the particles of most beaches, regardless of sediment grain size.

As a wave enters shallow water, the seabed below the wave interferes with the orbits, the circular movement of the water within the waveform. The now deformed and flattened orbits move forward and back at the seabed, with an increasingly forward motion. This action results in a net landward motion of water called the combined flow current, because it is a combination of oscillating and unidirectional water movement. It typically transports sediment toward the shoreline; but under some wave conditions, the reverse occurs, and net sediment transport is in the offshore direction.

The interference of the seabed with wave propagation also causes waves to refract as they approach the shore. Longshore currents caused by wave refraction range widely in their velocity, which depends on the size of the angle of approach and the size of the waves. They may exceed 100 cm/s under storm conditions. The combination of waves and longshore currents can move tons of sediment in a single day.

Sandbars that parallel a beach serve as modest, temporary traps for the water that is transported landward by waves. In an attempt to return to its original level, this water seeks out a path of least resistance and moves offshore through it. Typically this path is a low saddle on the sandbar—a rip channel. The current that runs through the channel is called a rip current and it, too, transports sediment.

The fourth mechanism for sediment transport in the beach environment is found in the swash zone, where the uprush and backwash of the breaking wave create swash currents that carry sediment across the foreshore. Depending on wave conditions, on the slope of the foreshore, and on sediment permeability, there may be a significant difference between the amount of sediment carried up the beach and that carried off the beach. One of the most important factors is permeability, the ability of sediment to transmit a fluid; a coarse and therefore permeable beach will have a greater uprush than backwash, because water percolates into the sediment and returns below the swash zone through the permeable medium.

Beach Cycles

The changes that occur on beaches can take place in an interval as short as the time between individual waves striking the shore. Or they can be subtle long-term changes, taking place over several decades. Likewise, the degree of change can be slight or it can be as dramatic as the extreme erosion caused by hurricanes. A level of predictability is possible, however, because these changes tend to come in cycles that respond to the weather's cycles.

Beach processes and their interactions with beach sediment can be considered in two distinctly different settings: fair-weather, low-energy, accretional beaches or foul-weather, high-energy, erosive beaches. At most beaches, fair-weather conditions include a swell wave with a low wave height (generally <1 m) and a period of 8 to 12 s. Locally generated sea waves may be superimposed on waves of a similar or smaller height with a period of 3 to 6 s. The sum of these low-energy conditions produces an accretional beach: The predominant condition is deposition of sediment or stability; erosion is absent or limited.

Accretional beaches have wide, well-developed backbeaches and relatively narrow steep foreshores. They tend to be reflective. The nearshore sandbars are generally well formed, and their relief is relatively high. Beaches of this type are a common configuration on most coasts with a low wave climate, for example, the coasts of the Gulf of Mexico and much of the Atlantic coast of the United States.

The alternate setting produces erosive, or storm beaches. Storms, although short in duration, are the dominant physical process along the vast

Beach erosion has toppled these homes on Nantucket Island off the coast of Massachusetts.

majority of coasts. During a typical storm, there is an increase in wind-wave size until the predominant wave is relatively steep. Sediment is entrained by these waves; and the accompanying currents readily carry it away, both offshore and alongshore. Additionally, swash energy is high and the resultant uprush and backwash are more extensive than during fair-weather conditions.

The removal of sediment from the beach produces an erosional profile, or storm profile. The backbeach is narrow or absent, and most of all of the beach is in the foreshore zone. A poststorm beach typically contains a ridge and runnel in the lower foreshore. The nearshore sandbars and related troughs move 10 to 20 m offshore and display less relief. The beach itself is covered with a veneer of high-density minerals accumulated as a lag deposit, that is, material left behind when the less dense components of the sand are removed by storm waves. This lag deposit of heavy particles is really a placer deposit, formed in much the same fashion as gold particles are concentrated in streams.

The storm beach is a temporary condition. In the absence of successive storms, a recovery period begins, characterized by the return to low-energy wave conditions. As swell and small wind waves resume, landward transport of sediment returns the nearshore sandbars to their original positions and configurations. More noticeable is the shoreward migration of the intertidal ridge as a result of flood-tide washover. The formerly somewhat symmetrical bar becomes asymmetrical, with the steep side landward. Over a period that ranges from two weeks to a few months, the migrating ridge rebuilds the beach, producing an accretional profile that resembles the prestorm setting. The beach again becomes reflective.

The recovery process can be interrupted by storms that return the beach to the erosional condition. The closely spaced storms of the winter season lead to what is commonly called a winter beach profile—essentially a storm profile. The frequent and severe storms along the west coast of the United States can remove an entire beach, exposing its bedrock bench (also called the wave-cut or the abrasion platform). This barren condition may persist for several months until the lower wave energy conditions of the spring and summer, when the sediment returns to the beach and eventually restores its accretional profile. Good examples of beaches that alternate between accretional and storm profiles are found along the central Oregon coast and near La Jolla, California.

COASTAL DUNES

Sand dunes are an important part of many coastal areas. They are large piles of sand that have accumulated by processes—and in shapes and patterns similar to those of the dunes found in inland deserts. The prerequisites for coastal and inland dunes are the same: a supply of sediment and the wind to move it. On the coast, the wind is seldom a limiting factor, whereas the sediment supply may be inadequate. In areas of abundant sediment, however, dunes exceed 100 m in elevation.

Coastal dunes are not restricted to barriers, although nearly all barrier islands have at least small dunes. And coasts without barriers can have very large dune fields. Particularly good examples of mainland dune fields are those found along the southern coast of Oregon, where the dunes extend 4 or 5 kilometers inland from the coast, and the shore along the southwestern part of Lake Michigan, where dunes over 100 m high have developed. Among the largest dunes in the world are those on the coast of Namibia in southwestern Africa, which also has no barriers.

Dune Formation and Distribution

Any coast where sand accumulates has the potential to develop dunes— prevailing winds or diurnal sea breezes provide the transport mechanism. Most coastal areas have prevailing winds with some onshore component

Small sand shadows have formed in the lee of shells scattered across a wind-blown backbeach. The location of the sand shadows tells us that the wind was blowing toward the observer.

that carries sand inland from the beach. But along the southwest coast of Australia near Perth, and on other coasts as well, the diurnal sea breeze is the dominant wind.

The dry part of any beach is subject to wind transport. The backbeach, which is especially susceptible because it is rarely wet, often shows various signs of wind transport, including sand shadows and gravel lag concentrates. The sand shadows indicate the most recent wind direction and may show scour around a shell or pebble. A gravel lag deposit appears when the wind blows the fine sand away and leaves the larger particles, which cannot be easily transported. After an extended period of wind erosion, the large particles become concentrated and actually form a pavementlike surface that inhibits further wind erosion. It is called desert armor because of its importance in limiting wind erosion.

Much of the wind-blown beach sand accumulates just landward of the active backbeach. Any type of obstruction, including bedrock cliffs, vegetation, existing dunes, or human construction such as buildings or seawalls, can stop further transport. Once the initiation of wind-driven sediment accumulation begins, it will continue until there is a change in condition, for example, the loss of the sediment supply or the destruction of the stabilizing factor, such as vegetation.

Vegetation is one of the best and most widespread facilitators of dune development, and any type of plant can serve as the focus for anchoring

Under the proper conditions, opportunistic plants may colonize dunes within a few months. Their continued growth stabilizes the dunes and protects the coastline against erosion.

wind-driven sediment. The relatively inactive backbeach is commonly covered with opportunistic plants, such as beach grasses and beach morning glory. Initially, small piles of sand accumulate around isolated plants on this part of the beach. In a matter of months, the piles of sand will increase in size.

Although vegetation supplies the best foundation for dune development, any sizable obstacle can initiate the process. Regardless of the impetus, small dunes or coppice mounds are eventually produced. But such small incipient dunes are quite vulnerable; even a modest storm can destroy them. The fragility of small dunes is one of the reasons why so much attention is now being paid to preserving the vegetation on the backbeach and at the base of dunes.

A dune ridge—a linear arrangement of dunes, one dune wide—is the typical configuration of dunes just landward of the beach. It is called the foredune ridge because of its location seaward or in front of the barrier or mainland. Many coasts contain numerous parallel dune ridges, each of which has formed immediately landward of an active beach as a foredune. The presence of several dune ridges marks a portion of the coast as one with a history of growth or progradation toward the sea. This is the optimal condition for any coast because it indicates an overall lack of erosion.

BEACHES, DUNES, AND BARRIERS

Some barrier islands contain a complicated assortment of dune ridge arrangements that show sets of ridges at acute angles to one another. These angled ridges indicate periods of erosion separated by periods of dune accumulation and barrier progradation.

Not all dunes are located adjacent to, or are associated with, the beach; some of these dunes migrate inland. Like those near the beach, inland coastal dunes are also dependent on an abundant sand supply and a means for moving and accumulating it. In most instances, one of two factors leads to landward dune movement—a huge sediment supply or an absence of stabilizing vegetation. An excellent example of an extreme abundance of sand is found along the coast of southern Oregon. Here the strong winds off the Pacific, acting on great amounts of sand, have produced huge mobile dunes that extend 4 to 5 km inland from the coast. These coastal dunes

Dunes migrate landward over a former pine forest at Umpqua Scenic Dunes in the Oregon Dunes National Recreational Area.

have buried forests of mature trees as they continue to migrate in a southerly direction parallel to the coast.

Desert conditions combined with extensive storm-washed tidal flats have produced an extensive, active dune complex in the central part of the Padre Island National Seashore in Texas. An additional assist was provided by extensive cattle grazing on the island during the late 1800s and early 1900s, when this area was part of the famous King Ranch. This portion of the Texas coast is characterized by considerable sediment accumulation and persistent onshore winds. As a result, Padre Island is extremely wide, and the mainland also is dominated by an extensive dune complex covering several hundred square kilometers.

The numerous, active dunes range from about a meter high on the landward side of Padre Island, near Laguna Madre, to several meters high in the central island. The smaller dunes on the landward part of the island are on supratidal flats where hurricane-associated storm surges a meter or more high can reach them. The storm surges flood the small dunes and flatten them by a combination of waves and currents. Then, after the surge subsides, the wind uses the available sand to reconstruct them. The dunes further toward the center of the island are less vulnerable to storm-related erosion, so they have grown larger.

Dune Dynamics

The very existence of dunes is testimony to the mobility of sand. Even though vegetation is an effective stabilizer of these sand accumulations, the waves of storm surges and the wind each have a profound effect on dune stability.

Although dunes are beyond the regular influence of waves, they are vulnerable to even the most modest storm surges, as illustrated by the Padre Island dunes mentioned earlier. This problem is particularly acute on erosive beaches because these areas have no backbeach to protect the dunes. Storm waves superimposed on an elevated water level produce swash—and, sometimes, direct wave attack—at the tops of the dunes. There the sand is easily washed away and carried both offshore and alongshore. Even if a dense cover of dune grass is present, it can be removed during a storm surge, thus leaving the root system hanging over the dune scarp. Poststorm recovery returns some or even all the sand to the beach. Under the right conditions, the rebuilding process of the dune will begin; but it can take many years to restore the losses incurred from even a single storm. Rising sea level exposes dunes to further erosion. As the sea level rises, tides and storms are even more likely to threaten the dunes with

Fingers of sand cascading down the surface of a coastal dune, pulled downward by gravity when the slope is steep enough.

BEACHES, DUNES, AND BARRIERS

The old town of Skagen at the northern tip of Denmark was buried in the shifting dunes. This painting depicts Skagen as it appeared in 1848.

wave attack—a scenario that occurred during much of the Holocene transgression.

The other major factor in dune dynamics is the wind, which can cause the migration of part or all of the dune. The winds that form the dune also can move it—sometimes great distances. Dune mobility is usually associated with an absence of vegetation. Foredunes, which are commonly vegetated, may experience local blowouts, an amphitheater-like excavation, after a change in the weather pattern reduces or eliminates their vegetative covering. Overgrazing also removes much of a dune's vegetation. Regardless of the cause, the result is the same—the sediment begins to move.

Blowover is the most common wind-driven process responsible for dune migration. Wind blowing in from the ocean (an onshore wind) simply carries the sand across the dune surface and permits it to move down the landward side by gravity. Blowover creates a relatively steep slope, generally with a maximum of about 30 degrees, which is called the angle of repose. All dunes, regardless of their location or direction of migration, are able to maintain this slope. When the angle increases, individual grains of sediment start falling down the steep slope and initiate a sediment slide—called grain flow—down the slipface. Oversteepening of the slope causes an instability that moves large numbers of grains down the slope in an avalanche fashion. Anyone walking down the face of a steep dune can start one of these avalanches underfoot. The combination of these wind-, wave-, and gravity-driven processes causes part or all of the dune system to migrate landward.

As discussed in the preceding section, the migration of large dunes pays little attention to trees, buildings, or any other obstacles in its path. As long as the dune is larger than the obstruction, it will move over it. Houses have been buried, and then many years later exhumed, by migrating dunes.

BARRIER ISLANDS

Barrier islands are only one type of coastal barrier; the others include barrier spits; barrier bars or barrier beaches; and barrier reefs. All except the barrier reefs are built of sand and other sediment that is supplied by waves, tides, and longshore currents. In this chapter, we will focus on barrier islands, because they are complex and include all of the environments also found in other coastal barriers. Other barrier types will be mentioned only in passing, or not at all.

Found worldwide, from the north slope of Alaska to the tropics of South America and Australia, barrier islands are elongate accumulations of sand that range from a few hundred meters to more than 100 km in length. Some are barely above high tide; others have dunes that rise 30 m above the sea. These wave-dominated and mixed-energy depositional systems constitute approximately 12 to 15 percent of the Earth's outer coastline.

Barrier Origins

Barrier islands develop in any geologic and tectonic setting that has a plentiful supply of sediment, agents to transport it, and a site where it can accumulate. Consequently, the best-developed and most extensive barrier islands are found on stable, trailing edge coasts. In these settings, sediment is abundant, low-to-moderate wave energy moves the sediment around but still allows it to accumulate, and a gently sloping continental shelf provides an adequate site for full development. From Narragansett Bay in Rhode Island to Miami, the Atlantic coast of the United States, a classic example of an Amero-trailing edge coast, is essentially one continuous barrier island system.

Marginal seas that have the three required ingredients—sediment, transport agents, accumulation site—also develop barrier islands. The coast along the Gulf of Mexico is an example of this setting.

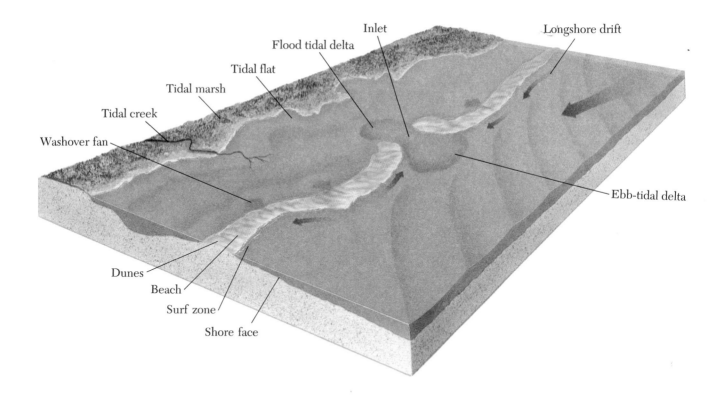

The major elements of a barrier island complex.

Tectonically active leading edge coasts, however, are unlikely places for well-developed barriers. Even when sufficient amounts of sediment are available, the energetic waves along these coasts tend to disperse it. The steep shores of these irregular coasts have little room for barriers. As a result, the only barriers commonly found along tectonically active coasts are short barrier spits. Along the Pacific coast of the United States these spits are numerous; there are nearly a hundred that are at least 1 km long.

Present-day, active barrier islands all date from the Holocene Epoch. Therefore, in geologic time they are quite young—nearly all are less than 7000 years old and many were formed less than 3000 years ago. Determining these dates is a rather straightforward process: Shells, woods, and other carbon-bearing material in the sediment contain radioactive carbon (C-14). This isotope continues to decay at a constant rate after the death of the living organism that incorporated it, so a count of the remaining C-14 atoms provides a good estimate of the age of any biological material between 500 and 30,000 years old.

Siletz Spit on the Oregon coast is an example of a barrier spit attached to a headland; it is a typical feature of a converging coast.

Unlike carbon dating, an accurate determination of how all types of sandy barriers formed is not so easy. Three origins for these barriers were proposed by geologists in the latter half of the nineteenth century. The first theory suggested that waves caused sediment to accumulate in an upward-shoaling fashion that eventually led to a supratidal sandbar. The sandbar then continued to accumulate sediment, became vegetated, stabilized, and formed a barrier island. The second theory advocated that a depressed strip of land between dune ridges and beach ridges was drowned by the rising sea level, thereby leaving the beach ridge separated from the mainland by a narrow and shallow water body; the beach ridge eventually developed into a barrier island. The third theory of origin proposed that after a spit formed at a headland it was eventually breached at one or more places to form a barrier island with tidal inlets. This mode of occurrence is well documented in some parts of the world such as the west coast of the United States and the east coast of Australia. However, because it requires a headland source for sediment, the barriers of the Gulf of Mexico and most of the Atlantic coast could not have formed in this manner.

Aerial photography has allowed us to document the formation of numerous very young barrier islands during historical time. All these young islands have developed essentially according to the scheme of upward-shoaling through wave action, and none has developed as the result of the drowning of a coastal beach-dune ridge system. It can be argued, of course, that sea level rise has been too slow for the drowning between the ridges to have occurred in visually recordable history. The key here is that a slow rise

BEACHES, DUNES, AND BARRIERS

in sea level will encourage either destruction or landward migration of the ridges, not their separation by drowning.

As noted earlier, the rise in sea level since the last period of widespread glaciation—or the Holocene transgression, as it is called—was very rapid for several thousand years. It is likely that this rapid rise did not allow a stable shoreline to exist long enough to develop barrier islands under any of the aforementioned theories. But about 6500 to 7000 years ago, the rate of sea level rise slowed dramatically. Shoreline positions became more stable and waves molded the coasts. These conditions fostered the development of barrier islands and allowed them to persist—some slowly migrated landward as sea level rose and inundated the old shoreline, and others just grew larger in the location where they had been formed. None seems to have arisen by drowning of the low-lying areas behind the beach ridges.

rejects 2nd theory

Barrier Island Components

Beaches and dunes are the environments most commonly associated with barrier islands and the seashore in general, but other components of the barrier system are important and often cover extensive areas. On the landward side of the barrier island, a large area exists between the dunes and the water line. Most of this backbarrier area is rather flat and only slightly above sea level. With little change in profile, the backbarrier grades into a marsh and eventually to a tidal flat that lies along the margin of the coastal bay, commonly a lagoon, separating the island from the mainland.

Washover Fans

The back island area, marsh, and tidal flat, are distinct environments with their own unique vegetation and elevation relative to mean sea level. Nevertheless, virtually all the sediment on which these separate environments have developed was originally deposited by one mechanism and accumulated in one type of sediment body—the washover fan.

A washover fan is a fan-shaped accumulation of sand and shell that is deposited in a thin layer during intense storm conditions when part or all of the beach-dune system is overtopped or breached by incoming waves and storm surges. On some barrier islands, individual fans coalesce to form a washover apron. On narrow barrier islands or under conditions of very high energy, the fans may extend completely across the island and into the adjacent coastal bay.

The sediment accumulations that represent a single storm may be as much as 1 m thick but more commonly are only a few tens of centimeters thick. Composed primarily of sand with some shell debris, they are generally well stratified in near-horizontal layers.

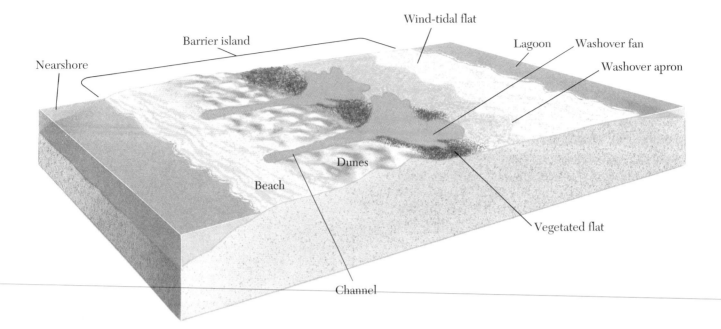

Nearshore

Barrier island

Wind-tidal flat

Lagoon   Washover fan

Washover apron

Dunes

Beach

Vegetated flat

Channel

Washover fans form when the dunes on a barrier island are overtopped or breached by storm waves. Adjacent fans may overlap to form a washover apron, a layer of sediment that may extend into the tidal flat at the back of the island, or in some cases even beyond.

The postdepositional phase of a washover fan is complex. The vegetation in the backbarrier is resilient. Despite partial, or even complete, burial, the terrestrial or marsh grasses living there sprout through the overlying fan sediment in a matter of months or even weeks. Occasionally, very thick fans will completely destroy the underlying plant community; however, the opportunistic barrier island vegetation commonly recolonizes the backbarrier in less than a year.

The upward penetration by partially buried plants and the downward penetration by surface plants disrupt the typically well-developed stratification of fan sediments. The part of the fan that extends into or below the intertidal zone is rapidly occupied by a broad spectrum of marine organisms, including marsh grasses, burrowing worms, and mollusks. Bioturbation by these organisms also destroys the stratification signature of the washover fan and contributes a small amount of fine sediment through the accumulation of fecal pellets. Additionally, the tides and waves act upon the distal end of the fan, further reworking its layers. All of these processes transform the package of sediment that was deposited as a washover fan into environments such as tidal flats, marshes, wind tidal flats, and back island grassy areas. Nevertheless, despite these transformations, the general shape of each washover fan is still detectable in aerial photographs.

Lagoons

Long, shallow lagoons form wherever barrier islands or sandbars separate a section of the ocean from the mainland. Lagoons are characterized by an absence of freshwater runoff and by a lack of tidal flux. These conditions prevent sediment flow from these areas; in addition to elevating the salinity of the water within the lagoon.

Sediment is primarily introduced into backbarrier lagoons from the barrier side of the water body through two mechanisms: washover during storms and blowover from onshore winds. Both mechanisms carry beach and dune sediments into the lagoon, but the overall rate of sediment accumulation in lagoons tends to be slow and sporadic.

The biology of lagoons tends to typify that of many hypersaline environments—few species but large populations. Small bivalves, worms, and shorebirds are abundant; and small fish thrive in the protected waters.

Elevated salinities in lagoons can reach extreme levels during dry seasons, especially in arid climates. High salinities can result in precipitation of evaporite minerals; gypsum forms at about 200 parts per thousand (nearly six times the concentration of normal seawater) and halite (salt) forms at about 300 parts per thousand. Calcium carbonate may also be precipitated in lagoons. All of these minerals can be found locally, in lagoons such as Laguna Madre, Texas, behind Padre Island and to the south in similar settings along the coast of Mexico.

The lagoon environment provides a sharp contrast to the estuarine environment. The nature of their sediments and their rates of accumulation are markedly different. Whereas estuaries show variation in salinity due to freshwater runoff and tidal flux, in lagoons these variations in salinity are more extreme and are due to seasonal fluctuations in precipitation (rain) and evaporation.

TIDAL INLETS

Barrier islands generally are breached at various points by tidal inlets, which link the open marine environment and the coastal environments landward of the barrier islands. Like beaches, tidal inlets are dynamic parts of the barrier island system and range widely in size, stability, and water flux. They owe their origin to a variety of circumstances, although storms and human activities are the most important factors.

Tidal Deltas

There are three major parts to a tidal inlet: the channel, or inlet throat, through which the tidal flux passes; the ebb tidal delta; and the flood tidal delta. The ebb and flood tidal deltas are accumulations of sand that vary greatly in size and shape and are located at the seaward and landward ends of the inlet channel, respectively. Like riverine deltas, they are built by the sediment deposition that accumulates when a current suddenly slows at the mouth of a channel, thereby losing its carrying capacity. In tidal inlets, of course, the short channel has a mouth at each end—because of the reversing tidal flow.

An inlet channel is generally as long as the width of the barrier island it bisects, but its width can range from tens of meters to a few kilometers. It is deepest—up to 30 m—at the narrowest part, usually near the middle of the barrier. A cross section of the channel reveals an asymmetrical shape, typically being deeper on the side that corresponds to the direction of the net longshore sediment transport (also called littoral drift). The amount of asymmetry is generally proportional to the stability of the inlet. The most stable inlets have nearly symmetrical channels, whereas those that migrate are more likely to be asymmetrical. And the more rapidly an inlet migrates, the more asymmetrical it will be.

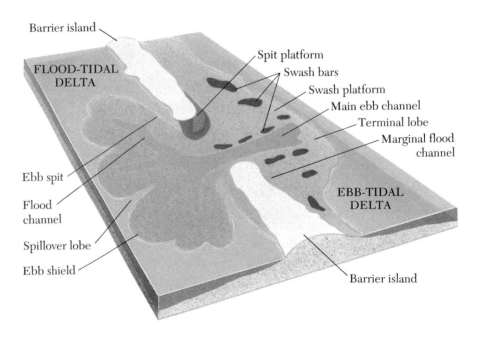

A flood-tidal delta and an ebb-tidal delta develop on opposite ends of a tidal inlet. Most of each tidal delta lies within the intertidal zone.

The stability of inlet channels is determined by the usual coastal agents. Tidal inlets, like the ebb tidal deltas to be discussed later, can be either tide dominated or wave dominated. Those that are tide dominated tend to be stable, whereas wave-dominated inlets either migrate alongshore or close completely.

We owe much of our terminology and understanding of the dynamics of tidal inlets, and especially tidal deltas, to Miles O. Hayes, a coastal geologist who spent considerable time at the University of Massachusetts and at the University of South Carolina. He and his students did most of the pioneering work on the morphodynamics of these important coastal elements in the late 1960s and early 1970s.

Because of their differences, flood tidal deltas and ebb tidal deltas will be considered separately. The flood tidal delta develops on the landward side of the inlet, so it is protected from significant wave influence. Tidal currents are the primary physical process to which it is subjected.

Flood tidal deltas generally are fan-shaped with a broad ramp that slopes up to the surface from the seaward end. This ramp, a continuation of the inlet channel, carries sediment to the tidal delta during flood tides.

The tidal range and the elevation of the flood delta strongly influence its outer shape. Along coasts where tidal range is about a meter or less, the

The multilobate form of the flood tidal delta at Chatham Inlet on Cape Cod, Massachusetts, as it appeared in the late 1960s.

shape of the flood delta tends to be multilobate, reflecting the delta's original configuration—especially those produced by storms. Because the tidal currents are generally unrestricted in their flow, they move back and forth over the delta without modifying it. Eventually many such flood deltas, typically only 2-3 m thick, become vegetated and further stabilized.

In areas where tidal range is 1.5 m or greater, the flood tidal deltas have a different appearance. The tidal currents in these areas are relatively strong and sediment transport is commonly high. The large tidal range causes the ebb tidal currents to be deflected around the lobes of the flood delta. This action, along with the regular addition of sediment across the flood ramp, produces a sand body with a relief of 1 to 2 m. The outer part of the tidal delta is smoothed by the ebbing currents that carry some sediment and deposit it in the form of small spits. Breaches can occur in the high part of the flood delta forming spillover lobes.

The depositional patterns and shapes of ebb tidal deltas vary more than those of flood deltas because ebb tidal deltas are exposed to the open ocean and its waves. This additional complication provides for a great variety in the amount and direction of the physical energy to which the tidal delta is subjected. Whenever a substantial supply of sediment is available, the tidal currents and their interactions with wave-generated processes control the size and shape of the ebb tidal delta. And the tidal currents are largely controlled by the tidal prism, which will be discussed later in this chapter. Suffice it to say that the greater the volume of water that must move through the inlet in a given time period, the faster it has to travel.

The three general types of ebb tidal deltas are tide dominated, mixed energy, and wave dominated. Tide-dominated ebb deltas tend to protrude

The three major categories of tidal delta. In a tide-dominated delta, the tidal currents moving in and out of the inlet shape the sediment bodies perpendicular to the barrier, whereas in a mixed energy delta, the sediment bodies have been smoothed by the waves to follow a curved seaward boundary. A wave-dominated delta is also smooth, but it is small as well.

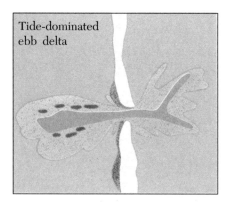

Tide-dominated ebb delta

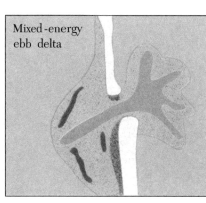

Mixed-energy ebb delta

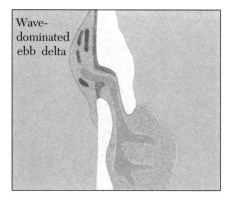

Wave-dominated ebb delta

into the sea at essentially right angles and have well-developed sandbars along their margins. Mixed-energy ebb deltas have a generally smoothed and curved outer sandbar, due to wave influence, whereas wave-dominated ebb deltas are small and have very little accumulated sediment.

As noted earlier, the tidal prism is the key factor that allows any of these types of ebb deltas to develop within a rather broad spectrum of tidal ranges. For example, although the tidal range on the western Gulf Coast of Florida is less than 1 m, all three types of ebb tidal deltas have developed there. Some of the inlets along that coast serve very large bays with large prisms, whereas others interact with only small bays, or share the prism of a bay with other inlets. Those with large prisms have tide-dominated deltas; the ebb tidal deltas of those with small prisms are either mixed energy or wave dominated.

Storms and Inlet Dynamics

Tidal inlets are affected by agents other than the tidal currents. These other agents of change are storms, tidal prism, and littoral sediment transport. Each is important and worthy of consideration, but we will consider storms first because they bring about the most rapid changes.

Many tidal inlets have originated during intense storms. Low and narrow barrier islands are vulnerable to storm-induced washover, which can excavate a channel and eventually produce a permanent tidal inlet. Several examples of this process are found on the peninsular Gulf Coast of Florida, an area where both hurricanes and low barrier islands are common. The most severe hurricanes recorded on this coast occurred in 1884 and 1921; both breached barriers to form permanent inlets. Johns Pass, which separates the southern end of Sand Key from Treasure Island near St. Petersburg, was formed by the hurricane of 1884 and has remained a relatively stable inlet—although its present shape has been modified by construction and some dredging. Redfish Pass, near Fort Myers, and Hurricane Pass, near Clearwater, were both formed during the hurricane of 1921. Both inlets are naturally stabilized and carry rather large tidal prisms.

The most well documented formations and development of a storm-generated tidal inlet took place as a result of Hurricane Elena, which occurred on Labor Day weekend in 1985. This modest storm did not make landfall along the peninsular Gulf Coast of Florida, but it passed close enough to influence the coast. Wave size peaked at a height of 2.5 m and a period of 13 s, with a storm surge increasing the water level about 1.5 m.

In 1989 Hurricane Hugo changed the landscape of Pawleys Island, cutting a new inlet from the intercoastal waterway to the ocean. The inlet was artificially closed a few months later.

This surge was enough to produce washover in several areas along the coast, particularly the narrow northern end of Caladesi Island near Clearwater. That portion of the island was normally only about 1 m above high tide and less than 50 m wide. A huge washover fan was produced, extending hundreds of meters into St. Joseph Sound behind the barrier. Initially, only a small tidal flow was able to pass through the washover channel, but it grew rapidly. Within three months, the channel was 1 m deep and about 30 m wide. By the end of one year, it was over 2 m deep and nearly 80 m wide but asymmetrical toward the north. After remaining open for 6 years, the inlet experienced a decrease in its tidal prism due to capture by an adjacent inlet, and it closed late in 1991.

Storms can also open inlets from the landward side of the barrier in a seaward direction. A very high storm surge can cause a bay to fill far above its normal capacity, up to several meters above the normal high tide line. Low places in barriers, especially formerly closed inlets, provide a path of least resistance for the escape of some of this water. A good example of this process occurred at Corpus Christi Pass near the boundary between Mustang and Padre islands on the Texas coast. The inlet had been closed for

BEACHES, DUNES, AND BARRIERS

several years because of sediment, which filled more than half the length of the old inlet. In 1967 Hurricane Beulah flushed the sediment into the Gulf of Mexico, producing a straight inlet with a channel nearly 2 m deep. However, in less than 10 years, the channel was filled again, essentially to its prestorm level.

Storm waves can also indirectly bring about important changes to inlets. Their high energy leads to the transport of great quantities of sediment by the longshore transport system, which then dumps the sediment in inlet channels. Dredging is commonly required to keep the inlets functioning.

Another storm-related phenomenon is the erosion of the ebb tidal delta. This sediment loss can be a serious problem for small inlets. In 1985 Hurricane Elena removed the small ebb tidal delta at Dunedin Pass, near Clearwater, Florida. As a result, longshore currents had direct access to the inlet mouth, and sediment from the longshore transport system filled the inlet within about two years. Dunedin pass probably would have closed eventually without the storm, but Hurricane Elena accelerated the event.

The depth of an inlet can be restricted by the geology of the underlying strata. For example, bedrock resists further downcutting. Or the channel floor can become armored by pebbles, cobbles, or shells, which accumulate as lag deposits similar to those on the backbeach. Once the inlet floor is paved, or armored, with these large particles, further downcutting is not possible.

Tidal Prism

The controlling factor in determining the size and stability of the inlet is the interaction of waves and tides at the seaward end of the inlet. Aside from storm surges, wave conditions at any particular location do not change greatly, so we will focus our attention on the primary variable in the inlet processes—the tidal currents and, more specifically, the tidal prism. The tidal prism is the amount of water exchanged in a given area during the tidal cycle. It is defined as the product of the tidal range and the area of the backbarrier bays that are served by the inlet. Large tidal prisms produce stable inlets; small prisms lead to instability, migration, and sometimes closure of the inlet. Large prisms create strong currents that flow through inlets, essentially perpendicular to the coast. These strong currents erode the channel, increasing its size until an equilibrium condition between tidal flux and inlet size is achieved. As long as the inlet cross section is increasing or remains more or less constant, the inlet will not migrate.

When the tidal prism is large, excess, storm-generated sediment moving along the coast accumulates near the mouth of the inlet in the ebb tidal delta. Continuation of this increase in sediment causes the tidal delta to become progressively larger, and eventually some of the sediment bypasses around the mouth of the inlet. If the rate of sediment flux decreases again, the ebb delta decreases in size as long as the wave climate remains the same, with most of the sediment being bypassed around the inlet.

If the prism is small, the tidal currents are generally weak. In this situation, longshore currents will carry sediment along the barrier and deposit it as a spit at the end of the barrier adjacent to the inlet. The continued and rapid accumulation of sediment by this process will lead to the migration of the inlet, in some cases by more than a kilometer. These conditions can lead to the eventual closure of the inlet, the rate of closure being dependent on the rate of sediment supply.

In a few cases, a balance exists between the sediment supply and the tidal prism and allows the inlet to maintain essentially the same cross-sectional area as it migrates along the coast. Some inlets may move great distances. Midnight Pass near Sarasota, Florida, moved in this fashion for 3 km, mostly during the nineteenth century. Generally what happens under conditions of prolonged inlet migration is that the narrow spit that forms on the seaward side of the barrier is breached by a storm and the inlet then follows a new path.

Because the tidal prism exerts ultimate control on inlet stability, changes in the size of the tidal prism are significant. The tidal range generally does not change, but other factors can affect the size of the tidal prism at a given inlet. Most frequently, a decrease in the size of the prism is caused by human activity, such as construction, which modifies the area of the bay served by the inlet. When the overall area of the bay decreases, the prism decreases and the inlet serving the bay becomes unstable.

Natural changes in tidal prism also occur. If the bay served by the inlet is an estuary with a significant influx of sediment from numerous sources, the tidal prism will slowly decrease in size as the estuary fills in. The effects on the inlet will be very gradual, perhaps acting over several centuries. A more drastic natural change in inlet stability is caused by the opening or closing of adjacent tidal inlets. Many barrier island systems along trailing edge coasts have multiple inlets that connect through the backbarrier water bodies. In fact, some adjacent inlets share tidal prisms with one more flood dominated and the other more ebb dominated—what goes in during the flood cycle of a particular inlet does not necessarily come out during the subsequent ebb cycle.

If a storm opens a new inlet on a barrier island, the tidal prism carried by that inlet must come from somewhere. In fact, it is captured from the previously existing adjacent inlet or inlets. In the event that the amount of water exchanged by the new inlet is sufficient to maintain a channel, the inlet will be stable and persist. However, most of these storm-generated inlets tend to have only small prisms and are commonly unstable and ephemeral. For example, a hurricane in Texas during the 1980s cut more than 50 channels through Padre Island, but none remained open for more than a year.

The opposite condition can also occur. When an inlet is closed, the prism it was carrying is transferred to an adjacent inlet. Most inlets close gradually, so the transfer is slow and unnoticed. However, if an inlet is closed by a storm, the transfer of tidal prism is abrupt and may cause the inlet to expand rapidly.

This inlet at Dunedin Pass on the Florida Gulf coast recently closed as a result of excessive littoral drift and a small tidal prism.

Barrier Island Dynamics

Viewing barrier islands in the third dimension through stratigraphy has contributed to an understanding of their dynamics. One of the first papers on this topic appeared in 1934 and dealt with the New Jersey coast. But investigations of barrier island stratigraphy really began in earnest with the study of Galveston Island, Texas, in the mid-1950s. Because of its proximity to Houston, where many major oil companies were located, it was a likely place to begin. The next detailed investigation was farther down the Texas coast, at Padre Island. Since the 1950s, there has been an explosion of research into the nature and morphodynamics of barrier islands, initially, by oil companies to develop oil exploration strategies but more recently, by the scientific and engineering communities for the purposes of coastal management.

Barrier Island Types

The interaction of the various coastal processes with the available sediment determines the shape of barrier islands. Indeed, whether barrier islands develop at all depends on the predominance of waves over tides. Regardless of the specific origin of barrier islands, waves and wave-generated currents must be present to produce the linear accumulation of sediments that makes up these coastal sediment bodies. When tides become dominant over waves, barrier islands give way to tidal flats and coastal marshes. These conditions exist in the corner of the German Bight on the North Sea coast of Europe and also on the open coast of the Florida peninsula north of the Tampa Bay area and near the Everglades.

Barrier islands generally assume one of two forms. One is long and narrow and derives its shape from distinctly wave-dominated conditions. They are likely to be transgressive in nature. However, when abundant sediment is available, multiple parallel dune ridges develop as they prograde. The Outer Banks of North Carolina and the barrier islands of the Texas coast are two examples of the wave-dominated form.

Tidal inlets associated with the wave-dominated barriers tend to be widely spaced, small, and unstable. Inlets migrate by means of spit development at the end of the barriers. Over a long time period, the direction of littoral drift repeatedly reverses, so inlets move back and forth, filling in behind themselves as they migrate. Some close completely, and new ones are created during storms. These processes produce a barrier island stratigraphy that consists of a thin beach and dune accumulation overlying a thicker sequence of sediments, which was deposited as fill by the migrating inlets. The stratigraphy under the Outer Banks has been estimated to contain nearly 50 percent by volume of inlet-fill deposits.

The other form of barrier islands is relatively short, with one end much wider than the other. Barrier islands of this type are built up and maintained by a combination of wave- and tide-generated processes. Such mixed-energy barrier islands have been named drumstick barriers by Miles Hayes, because of their resemblance to a chicken drumstick. They have developed in the German Bight, the Georgia Bight, and the west Gulf Coast of Florida.

The asymmetric shape of these coastal elements is the result of the interaction of the inlet and the barrier island. In a mixed-energy inlet a modest, curved, ebb tidal delta protrudes into the sea. As waves approach this shallow accumulation of sediment, they are refracted around it, and this refraction pattern actually causes a local reversal in the longshore current and therefore in sediment transport. As a result, sediment is trapped at one end of the barrier island and the other end is starved. The island progrades at the updrift and adjacent to the tidal inlet and transgresses at the narrow, low-lying downdrift end.

The deposition pattern of drumstick barriers eventually leads to a shift in their orientation with respect to the adjacent shoreline. Caladesi Island on the Florida Gulf Coast has experienced a change in shoreline orientation of 15 degrees during the last century. A continuation of the pivoting

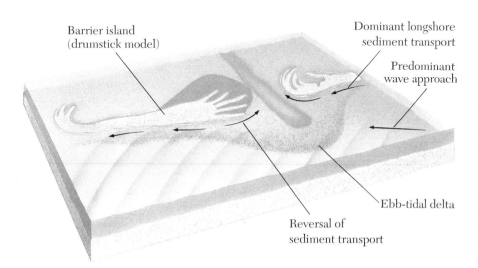

Barrier island
(drumstick model)

Dominant longshore
sediment transport

Predominant
wave approach

Reversal of
sediment transport

Ebb-tidal delta

Where a mixed-energy inlet creates a smooth tidal delta, the refraction of waves around the delta reverses the longshore current. The local reversal of current and, therefore, littoral drift, causes abundant sediment to accumulate on the updrift end of the island, which takes on a drumstick appearance.

processes produces barrier island shorelines that are markedly offset on each side of the tidal inlet. This downdrift offset, as it is called, is a unique feature of mixed-energy barrier island systems.

Transgression and Progradation

Storm-generated waves carry sediment from beaches and dunes to the back of a barrier island and deposit it as washover fans. If this deposition continues over an extended period of time, a landward displacement of the barrier occurs, conveyor-belt style. If the sea level rises, then barriers become even more vulnerable to this type of conveyor-belt movement. The motion of a landward-moving barrier island is called transgression—in its landward movement, the island "transgresses" over backbarrier environments. Barrier island transgression has occurred along the eastern shore of Virginia, on shores in the Georgia Bight of South Carolina and Georgia, and along many of the Gulf Coast barriers.

Evidence of barrier island transgression appears on the beach and in the surf zone. Peat, tree stumps, and layers of fine-grained organic mud all represent elements of a quiet, protected, backbarrier environment, such as a marsh or mangrove swamp. It is certain that these backbarrier features did not accumulate in a surf zone where they are now found. Their presence in sandy environments is proof that the beaches, dunes, and washovers are moving landward over older, backbarrier environments.

As the barrier transgresses, it loses some of the sand-sized sediment, leaving it behind in the nearshore or inner shelf. Unless more sand becomes available, largely through longshore transport, the barrier is destroyed before it reaches the mainland. If the barrier reaches the mainland, however, it merges with the mainland coast in much the same fashion that a ridge and runnel system migrates and welds to a beach during and after a storm. There is, of course, a major difference in scale and complexity.

The sea level is rising rapidly along the barrier coasts of the Mississippi Delta area, and conditions there demonstrate how future increases in sea level rise will affect barrier islands in other areas. A cooperative study of this area has been completed by the U.S. Geological Survey and the Louisiana Geological Survey. They have found that some of the barrier islands that developed on abandoned lobes of the Mississippi Delta are transgressing landward at rates of several meters per year as a result of their low elevations, the rapidly rising sea level, and frequent washovers during storms. They have also found that the barrier islands are becoming smaller

The presence of peat deposits and tree stumps is proof of the transgression of this barrier island on the Georgia coast. The beach has migrated over the back-barrier marsh, exposing it to wave action.

as they transgress. These changes are consistent with our current understanding of barrier island dynamics, and more of the same can be expected elsewhere if the sea level continues to rise at an accelerated rate.

Addition of sediment can also cause barriers to prograde, that is, to grow in a seaward direction. This condition is different from transgression in that the barrier island as a whole does not move. Instead, the addition of sediment causes the development of multiple beach-dune systems, and the open water shoreline actually moves seaward while the landward backbarrier shoreline remains in place. Sea level can rise, fall, or remain static, but the important factor in progradation is the delivery to the beach and dune system of more sediment than can be carried away either alongshore or over the barrier. Examples of this condition are present in parts of Australia and on the Nayarit coast of Mexico.

An individual barrier island can actually experience transgression and progradation at the same time. If the sediment reaching the barrier island is not uniformly delivered along its shores by waves and longshore currents, then most of the sediment can be trapped at one end of the barrier island while the other end suffers a deficit. The result of these simultaneous actions is progradation at the end of the island receiving sediment and transgression at the sediment-starved end.

Barrier islands are at once complicated, dynamic, and fragile. Those we have off our shores today are geologically quite young. If the forecast for a rapid rise in the global sea level is accurate, they may not become much older.

7

Rocky Coasts

Rocky coasts, with their spectacular cliffs and huge waves, extend over 75 percent of the world's continental and island margins. For example, much of the northern half of North America, southern and eastern Australia, the western coasts of Asia, the northern Mediterranean, and most of the islands of the world are dominated by rocky coasts. The age and nature of the rocks vary widely, but the wave climate is always energetic.

ORIGINS OF ROCKY COASTS

Rocky coasts are commonly tectonically-active convergent coasts that produce a high-relief border. Because they are formed on continental plate margins, under which an oceanic plate is descending, virtually no continental shelf is present. The western edges of North and South America are excellent examples of this type of coast.

The geology of the American convergent coasts is a combination of various types of rocks in structurally complicated settings of faults, folds, and igneous intrusions and extrusions.

Rocky bluffs near Otter Cliffs, Acadia National Park, Maine.

Steep slopes dominate the typography, both above and below the shoreline. The adjacent Pacific Ocean basin allows large waves—the typical period is 10 to 14 s—to develop and reach the coast unhindered by shallow water. However, the extent to which the cliffs are directly produced by the waves is undetermined.

On other, tectonically unrelated, cliffed coasts, various sedimentary strata are horizontal or dip at low angles. The adjacent continental shelf is wide, with a gentle slope. But here, also, the waves are large because of great fetch. Examples of these broad, cliffed coasts are found on the southern and eastern coasts of Australia and the western coast of New Zealand, many areas of Great Britain, and parts of the European mainland. Bluffs, well over 50 m high in some places, extend for hundreds of kilometers along the Nullarbor Plain in western Australia. These coasts have been exposed to wave attack for many thousands of years and are spectacular and rugged.

The rocky coast along a fjord at Kenai Fjords National Park, Alaska, was carved by a glacier.

ROCKY COASTS

Irregular bluffs of glacial drift show erosion on the coast at Gay Head, Martha's Vineyard, Massachusetts.

Pleistocene glaciers have also had a hand in producing cliffed coasts. The moving ice masses gouged out steep valleys, which were subsequently drowned as the sea level rose. Although their profiles are similar, some are rocky and others are not. The spectacular fjords of Scandinavia, Greenland, and other high-latitude areas are drowned valleys of this type. The low rocky coasts that dominate the ancient shield areas along the northern coast of Canada and parts of Scandinavia also began as glacier-made valleys.

Still other cliffed coasts are formed of glacial drift—sediment deposited by glaciers beneath and at the margins of the ice. The drift is over 100 m thick in some places and includes nearly any type of material from stiff muds to sand and gravel. Some of it is well layered and some is massive, with essentially no internal organization. The accumulations known as end moraines tend to be linear and thick. When these end moraines meet the sea, the waves sculpt steep bluffs. One of the most spectacular glacial-drift coasts is at the Cape Cod National Seashore in Massachusetts. Much of the outer portion of Cape Cod, which trends north–south along the Atlantic Ocean, and the coasts of Martha's Vineyard and Nantucket are glacial end moraines. Similar but less impressive moraine bluffs are present around large sections of the Great Lakes coast, such as the Scarborough Bluffs west of Toronto and the southeastern part of Lake Michigan.

Another variety of rocky, and commonly cliffed, coast is associated with areas where the continental shelf and adjacent coast are dominated by skeletal shell debris. Such coasts are found in the low-latitude Caribbean and Mediterranean and along the high-latitude southern coasts of the Australian continent and South Africa. In tropical areas the shells of dead marine organisms yield the mineral calcium carbonate at a high rate. This mineral is the only sediment available, as there is virtually no terrigenous, or land-derived, sediment present. The continental shelf of southern Australia, bordering the cold Southern Ocean, is hardly a tropical system, but it also supports shell life and no major rivers exist to supply it with terrigenous sediments.

A similar type of rocky coast has been constructed from the abundant carbonate sediment in these contrasting areas. In the Pleistocene, onshore winds blew carbonate sediment landward, where it accumulated in wide beaches and dunes. In a process called lithification, the calcium carbonate grains are welded together by a cement created as ocean spray or percolating ground water reacts with the calcium carbonate. The evaporation of the regularly wetted surfaces in arid climates enhances the lithification of the sediments. The rapid cementation converts the dunes to a rock called eolianite. Today the extensive eolianite deposits accumulated during the Pleistocene Epoch form rocky and cliffed coasts of Bermuda, many Caribbean islands, and the Yucatan coast of Mexico display spectacular large-scale cross-stratification and soil horizons, which indicates that periods of weathering alternated with periods of sediment accumulation and lithification.

PROCESSES

The processes that operate along rocky coasts are the same as those occurring along any other type of coast, but there are noteworthy differences in their effect. Waves and tides operate differently on rocky coasts than on gently sloping coasts. Wind and storms, freezing temperatures, and rainfall influence rock weathering. Biological and chemical processes, although important, affect the coasts at a much slower rate and on a smaller scale than the physical processes do.

Physical Processes

Wave energy is high on rocky coasts, at least relative to the adjacent beaches. The size of the waves that reach the coast is related to the near-

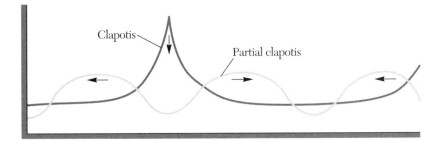

Clapotis and partial clapotis are produced by waves reflecting from a vertical cliff.

shore bathymetry and the resulting refraction patterns assumed by the waves. Aerial photographs clearly show waves bending around the headlands into the adjacent bays or inlets, where the sediments that make up the beaches accumulate. The wave energy is focused on the headlands and dispersed in the bays, so the headlands erode as the intervening bays fill up.

Many of the waves reaching rocky headlands, especially ones with a steep subtidal slope, are reflected back from the cliff with little or no loss of energy. As a wave hits the base of a cliff, it reflects and reinforces the next incoming wave. The second wave instantaneously becomes a standing wave larger than the incoming waves. It appears as a vertical loop of water alternately rising and collapsing—a waveform called clapotis. Clapotis waves do not break, so their energy is not dissipated. If there is shallowing near the cliff, then some wave energy is dissipated and a partial clapotis is produced. These waves travel both landward and seaward from the collapsing clapotis.

When a wave hits the steep rock surface, pressurized air is trapped beneath it and provides a cushion for the "water hammer" that the wave represents. But the pressure itself works on the rock. From various measurements and models, investigators have concluded that the greatest pressure exerted by waves on a surface occurs at, or slightly above, the still water level. This conclusion is supported by numerous notches and eroded surfaces observed at this elevation on rocky cliffs.

One of the most important process of physical erosion is the abrasion of rocky coasts by waveborne particles. The rapid transport of large gravel particles across the rock surface produces considerable impact and abrasion. In the British Isles, where rocky coasts are widespread, researchers have attributed most of the cliff erosion to this mechanism. Unfortunately for this theory, there are numerous parts of the world, such as the eastern coast of New South Wales and much of Tasmania in Australia, where ero-

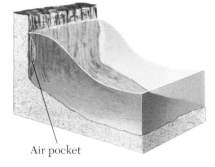

Air pocket

Waves that break against a vertical cliff will trap air, under pressure, for an instant.

sional features are widespread but large particles are absent. However, sand can also be an abrasive agent. Or virtually any hard particles transported in the energetic wave zone can act as erosional tools on a rocky coast.

Tidal range on rocky coasts varies greatly. Most of the Bay of Fundy, and the east margin of the Bay of St.-Malo in France, the two areas with the world's greatest tidal range, are bounded by a rocky coast. But the European coast of the Mediterranean Sea, with a tidal range of about a meter, also has a rocky coast. In isolation, tides play an indirect role on rocky coasts and exert a somewhat benign influence on them. However, the slope of the intertidal zone and the tidal range determine the size of the area affected by waves. Steep or vertical cliffs tend to have wave energy focused in a narrow intertidal area, whereas a gently sloping rocky coast spreads the wave energy over a relatively wide zone. The place on the coast where waves work the longest experiences the greatest wear. Macrotidal coasts are coasts on which wave energy is distributed over a relatively wide zone; on microtidal coasts, wave attack is focused on a narrow zone. The erosional notches in cliffs are testimony to this relationship. These notches are more common along microtidal coasts than along macrotidal coasts.

Storms do not appear to have important effects on rocky coasts. Whereas the profile of a sandy beach can change significantly during a storm, rock cliffs are not similarly affected. There are no erosional features above sea level that correspond to storm surge levels.

Other, more subtle physical processes contribute to change along rocky coasts. They are typically temperature dependent and involve freezing and thawing or changes in volume. When water freezes under some degree of confinement—cracks, crevasses, or joints—it can break rocks. Porous and permeable rocks are particularly vulnerable to frost damage. The water comes from three sources: groundwater, which percolates up from the land through the cliffs; rain; or waves and aerosols from the sea. Groundwater and rainfall are intermittent, occurring locally or at specific times, whereas seawater is always present throughout the shore zone. Although the freezing temperature of seawater is depressed nearly 2°C, freezing regularly occurs in high latitudes. The intertidal zone tends to be the area most affected, because freezing and thawing can occur with each tidal cycle. The stress caused by repeated freezings and thawings results in the physical deterioration of the rocks and leads to erosion. Rocks that are foliated, or thinly layered, like schist or sandstone, absorb the most water, and they are most susceptible. Massive rocks such as granite or basalt absorb less water and are relatively more stable.

Although more effective as an erosion agent through the freezing and thawing of water, evaporation and temperature change alone can cause the mineral grains and rock fragments to expand and contract slightly. For example, halite, common salt, commonly precipitates from evaporation or salt spray. This mineral also expands at more than twice the rate of granite, and salt crystals that precipitate in the surface layer of the granite rock along the shore can cause deterioration of the rock as they expand and contract through heating and cooling.

Biological Processes

Although not as rapid or as extensive as physical processes, biological and chemical reactions have important long-term effects on rocky coasts. Some biological and chemical effects are readily visible; others go unnoticed.

The obvious contributors to erosion are the plants and animals that bore into the rock surface. Three phyla and many species of microscopic boring algae make up the most effective and widespread group of bio-eroders of rocky surfaces. They range from supratidal levels to well below low tide, penetrating as much as 1 mm into the rock. The major factor in their effectiveness is their population density, up to a million per square centimeter! The effect of these organisms is related to porosity of the rock type. They are relatively scarce on hard granite but are very abundant and penetrate deepest into limestone and other porous carbonate rocks, such as on reefs. Rocks bored by algae are eroded rapidly, grain by grain, as the waves strike again and again.

Various bivalves, sponges, worms, and spiny echinoids are also important borers of rocky coasts. They typically penetrate several millimeters into the hard surface of the rock for shelter, weakening the rock and making it susceptible to wave attack, especially where the borings are closely spaced.

Other species occupy the surface of the rocks. The most common feeding style of the grazing animals—chitons, limpets, snails, some sea urchins, and seastars—is slow scraping of the surface for lichens, fungi, or algae. The boring algae are a favorite food source. The extensive weakening of the rock surface by algae, combined with the scraping of the grazing animal, produces slow but significant erosion.

Chemical Processes

Chemical weathering is produced by the slow chemical reactions that take place between the minerals that comprise the rocks and the local environment. Because these reactions are slow, they often go unnoticed by the casual observer, but they are important in coastal zone dynamics.

The two most important factors in chemical weathering are climate and rock composition. The primary agent for chemical weathering is water. Hot, humid regions, such as the tropics, experience considerable weathering, whereas arid deserts experience virtually none. Limestone rocks (calcium carbonate) are quite soluble in the ambient moisture, but quartz, the common constituent of sand, is nearly inert.

Oxidation is probably the most obvious and rapid of these chemical reactions, especially the oxidation of iron. There is a considerable amount of iron in many minerals, and the addition of oxygen from water changes ferrous oxide to ferric oxide—essentially producing rust.

The structure of many common minerals breaks down as a result of a reaction with water, or other compound, that is present in the surface water. For example, carbonic acid, which is present in small amounts in surface water and groundwater, can slowly dissolve limestone. Feldspar, a common constituent of granite and other widespread rock types, reacts with water to produce clay materials as a product of weathering. Most of the mud on the Earth's surface and within the lithosphere has a very high clay mineral content, most of which was produced by chemical weathering. The rates of these reactions range from centuries for the dissolution of limestones to millions of years for conversion of feldspar to clay minerals. The specific climatic conditions are very important in regulating these rates. Rapid chemical weathering occurs in the hot, humid Amazon River basin where soils are thick and the volume of mud carried by the river is high. On most rocky coasts, however, the rates of chemical weathering are too small to be measured within one's lifetime. Time is an important factor in the weathering process. Except for limestone, it takes thousands to millions of years to break down most rocks. In the meantime, the environment or even the location of the rock may have changed.

Rock building is generally not a factor in the dynamics of the coast, except where shells and other calcium carbonate grains are cemented together to form limestone rock. The combination of a humid tropical climate and nearly continuous exposure to saltwater ensures a high level of dissolved calcium level in local waters. The result is beachrock, a type of limestone formed at the shoreline from beach deposits. It commonly includes soda cans, beer bottles, and other artifacts that testify to its young age. Dunes may also solidify, forming eolianites—rocks formed from wind-deposited material. The rapid formation of such rock material along the shoreline area stabilizes the coast and resists wave and current action.

Lithified sand grains due to precipitation of calcium carbonate cement.

This rock photographed near a beach in San Mateo County, California, is perforated by the spherical hollows called Tafoni.

Chemical weathering produces interesting surface features and patterns on rocky coasts. One such feature comprises abundant, somewhat spherical hollows that characterize a range of rock types in many parts of the world. These hollows are called tafoni, a term that originated in the Mediterranean where this feature is quite common. Their actual mode of formation is unknown, but relationships with climate, particularly moisture, suggest that a chemical process plays a role. Honeycombs, a somewhat similar feature, commonly occur in association with tafoni. Honeycombing is caused by boring sea urchins or small-scale fracturing in layered rock—although it also occurs in apparently homogeneous rocks. Both tafoni and honeycombs are common in the intertidal zone.

Where the Work Is Done

Each of the physical, biological, and chemical processes that affect rocky coasts has a particular domain, but none is restricted to the portion of the coast that we can see; much of the activity has a subtidal influence. Certainly weathering is more pronounced above sea level, and freezing and thawing are far more important above low tide. But biological processes and wave influence are especially important below the water line of rocky coasts. It is the part of the rocky coast that receives most, and sometimes all, of the wave energy. Life is also abundant in the subtidal zone, with the same problems of food and shelter. Unfortunately, very little is known about the subtidal dynamics of these high-energy zones because the high energy makes it both difficult and dangerous to make observations or to place instruments there.

GEOMORPHOLOGY

The broad range in rock type, structural configuration, and wave climate ensures that, like fingerprints, no two reaches of rocky coast are alike. Some are steep and others are not, some are stable and others are not; the

Rempton Cliffs in Yorkshire, England are a good example of vertical cliffs. Note the people atop the cliffs for an indication of scale.

variety is endless. Some specific styles and landforms, however, do tend to develop along rocky coasts. Most noticeable, of course, are the steep cliffs or bluffs that dominate the landscape. These cliffs display a considerable range in height and steepness. Some are nearly vertical and rise over 100 m above the water surface. Examples include the White Cliffs of Dover in England, the cliffs on the southwest coast of Victoria in Australia, and the rocky coast near the mouth of the Columbia River in the state of Washington. Each coast is composed of different types of rock, but all have a comparable shape and scale.

At the other end of the range are the low rocky coasts that owe their origin to beach rock. The beach profile of these rocky areas is similar to that of a standard beach, with low, seaward-dipping layers and an elevation that tends to be intertidal or just above high tide. Many low latitude coasts, such as the Yucatan Peninsula and parts of the Persian Gulf, have extensive stretches of beach rock; in many places they are associated with coastal dunes that are also lithified.

Because of the broad range in scales and slopes, it is not practical to examine in detail the various classification schemes of rocky coasts. As long as the coast has some type of rocky or resistant material that produces a steep profile, it is classified as a rocky coast.

Vertical Features: Cliffs

The spectacular landscape of the rocky coast owes much of its character to the steep cliffs at or just landward of the shoreline. Most of the cliffs are nearly vertical and are composed of bedrock, although these features vary widely worldwide.

A survey of the distribution and size of cliffed coasts on a global scale shows a concentration of high elevations above sea level in the mid-latitudes, with relatively low cliffs appearing in the low and high latitudes. This discrepancy occurs because high wave energy is needed to produce the higher coastal elevations, and it is greatest in the mid-latitudes. In high latitudes there is a period of wave inactivity because the coast is covered with ice. In tropical areas the winds are less intense and offshore coral reefs tend to buffer wave energy. Another factor affecting the distribution and size of bluffs is glacial erosion, which has reduced the relief in the high latitudes. In the low latitudes, many rocky coasts are dominated by lithified dunes, which are limited in their elevation. Around the world, however, local geology and wave conditions have created numerous exceptions to these general patterns.

Geology is probably the most important consideration. Not only lithology but also the stratigraphy, surrounding structures, and the attitude of the rocks are important. Under the same environmental conditions, these various geologic attributes can produce a wide range of coastal configurations.

Massive rocks, such as granite, will erode in a uniform fashion because of the overall uniformity of the material. Layered sedimentary rocks will display some heterogeneity between layers and respond to various processes accordingly. The erosion of alternating layers of sandstone and shale, for example, produces an irregular cliff in which the softer, more rapidly eroding shale beds have receded between ledges of resistant sandstone. When the sedimentary layers are thick, they present another variation, depending on where the softer material is positioned relative to sea level. If the shale layer is in the zone of direct wave attack, the rate of erosion is much higher than if this zone were occupied by a resistant sandstone. Large overhanging ledges of sandstone can collapse, causing pulses of rapid cliff retreat. In addition, digging and boring creatures prefer the softer shales, which further weakens them and increases the rate of erosion.

Waves have eroded the softer layers of these sandstone cliffs on the Tasmanian coast of Australia.

Steep cliffs of unconsolidated sediments constitute much of the shoreline along Drake's Bay in Pt. Reyes National Seashore, California.

The steepness of cliffs and the processes to which they are subjected are also related to their geologic characteristics. When wave processes are distinctly dominant, the more resistant and homogeneous rock types tend to produce the steepest rock facies. But when weathering from spray and windborne gravel is important and waves are small, the bluffs slope more gently. Jointing or other fracturing in the rock enhances cliff erosion. These surface weaknesses leave their signature in irregular, steplike profiles on the eroded cliffs. Variation arises because some rocks are tilted from the horizontal; in fact, virtually all orientations are possible. Folding of the strata further complicates the picture.

Some coasts have very high, steep bluffs of unconsolidated sediments. Although they are not truly rocky coasts, they are included here because their morphology is similar. The bluffs are typically formed by the erosion of glacial sediments and older coastal deposits, such as dunes. They tend to be fronted by beaches that develop from the interaction between lag material, which is produced as the cliff erodes, and the action of the waves. The erosion rates for this type of cliffed coast are commonly high, although rates of cliff or bluff retreat can range from nearly zero up to many meters per year. The combination of resistant rock, such as quartzite, and low-to-modest wave energy produces no measurable retreat. (Quartzite is essentially inert chemically and is too hard for most boring organisms.) By contrast, recession rates of several meters per year can occur where eruptions place erodible volcanic rocks along the coast. The more common rate of cliff retreat ranges from about 1 mm per year in resistant homogeneous rocks to 1 to 2 m per year in soft shales and friable sandstones.

The presence of steep prominences along the coast does not preclude beaches; they occur at the foot of cliffs or in the relatively quiet wave conditions between headlands. These beaches display the same morphology and experience the same dynamic conditions that barrier beaches do. Along many rocky coasts, the presence of sandy beaches is ephemeral, coming and going with the seasons in step with wave energy conditions. Gravel beaches tend to persist in these settings and show changes only during intense storms. However, the presence or absence of a beach does not significantly influence the cliff other than to provide some protection from waves.

The different configurations of coastal cliffs and the adjacent nearshore area. Platforms may be essentially horizontal, either clearly visible above the high tide level or mostly absent from view at the high tide level; they may be completely absent.

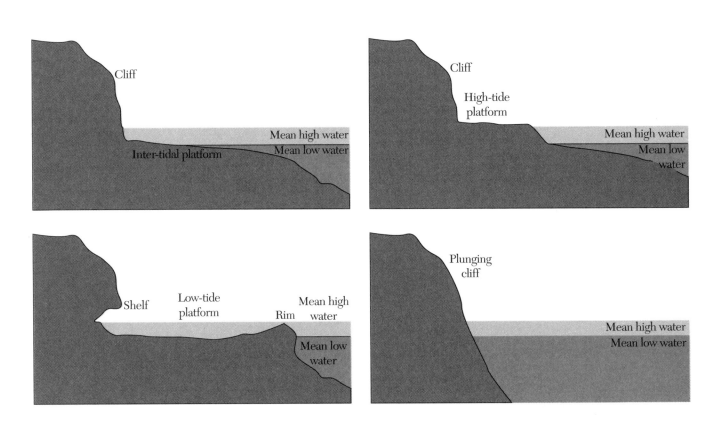

Horizontal Features: Platforms, Benches, and Terraces

Rocky coasts, especially those characterized by fairly high and steep cliffs, tend to have occasional flat, horizontal surfaces. These surfaces are primarily shaped by the long-term action of waves. In fact, they were formerly called wave-cut platforms, but geologists have recognized that other phenomena also contribute to their origin. The term shore platform is now used for these features, because it does not connote origin.

Bedrock is very resistant to erosion, but over an extended period of time waves can have an effect. For a significant amount of erosion to take place, however, the sea level must be stable enough to permit the same portion of the coast to be repeatedly exposed to the work of waves. In some parts of the world, such as Australia and New Zealand, this condition has been met because the sea level has been stable for approximately 6000 years.

This shore platform at Schooner Gulch, Mendicino State Park, California, was cut by waves across dipping sedimentary strata.

The origin of shore platforms is generally ascribed to a combination of wave action and a type of weathering called water-layer leveling produced by small pools of standing seawater left by the receding tide. The size and density of these shallow pools depend on rock type and structural conditions. Different proportions of chemical precipitates such as calcite and chert along joints and other fractures produce distinctive patterns in the linings of the pools. And this uneven relief of linings in turn creates other very shallow pools where further weathering can take place.

Subtidal platforms, or terraces, generally receive little attention because of their position below the visible coast. These wave-formed features show up in depths of only a few meters to about 10 m below sea level (depths up to which waves can have a considerable impact), depending on wave climate and rock type. Terraces are particularly common in less resistant rocks, such as limestone.

The eroded platforms of rocky coasts are typically horizontal, but some have seaward-sloping surfaces, with gradients of less than 5° to as much as 30°. Their position relative to sea level ranges from about mean tide to just above high tide. Sloping shore platforms can be partly attributed to rock type or inclination of strata, but many do not show this relationship. It has been suggested that a better correspondence exists between the gradient of the shore platform and the tidal range, the rationale being that as the tide rises and falls over a greater range, it provides a mechanism for spreading out the influence of waves on the shore. The greatest wave energy is directed at the seaward edge of the platform. Conversely, the least amount of wave energy is expended at high tide. Using data from several areas, investigators have constructed a plot of the tidal range and platform gradient that seems to show a trend supporting this relationship. However, considerable spread in the gradient is seen at several of the areas.

Areas of tectonic activity and places where the sea level has been much higher than it is now have terraces above sea level. They can vary widely in size and elevation. A few coastal areas even have several terraces. Along the California coast, tectonic activity associated with the plate boundary has produced multiple terraces tens of meters above the present sea level. The reverse can also occur—drowning terraces. These drowned terraces may be partially or completely buried by sediment, further obscuring them.

Stacks, Arches, and Other Erosional Remnants

Various parts of a particular rock differ in their ability to resist erosion. Differences in resistance are the result of structural weaknesses in the rock caused by fracturing, or the layering of various rock types, or different levels of cementation. The one feature that all stacks and arches share is some type of vertical character.

Sea stacks and arches are formed from the resistant rock that remains after less resistant surrounding material has been eroded. Known as differential erosion, the condition begins when waves attack the weakened areas of the shoreline bedrock, chiseling away at the fractures in the surface rock to form fingerlike indentations. As continued erosion separates the headlands from the mainland coast, isolated sea stacks of resistant rock emerge. Continued wave attack will cause the shoreline to retreat, further separating the sea stacks from the coast. For example, a volcanic neck composed of relatively resistant basalt can remain long after the surrounding volcano of tuff, ash, and other relatively soft material has been eroded by wave activity.

Isolated sea stacks are most common, but some locations contain numerous stacks. The Twelve Apostles on the Great Ocean Highway in western Victoria, Australia, is one of the most spectacular groupings. This family of sea stacks ranges in size and distance from the present coast and appears to form a cluster when viewed from almost any angle. The long coast of southeastern Australia is especially spectacular. Cliffs of Miocene limestone rise 50 to 60 m above the ocean for about 100 km. Jointing in the bedrock triggers the production of a wide variety of erosional remnants in the form of stacks, arches, and caves.

Sea arches are commonly associated with sea stacks and may, in many cases, be their precursors. As their name suggests, these arch configurations of rock are also the result of differential wave attack and variation in rock resistance. Arches are found in essentially the same geologic setting as sea stacks. The difference is that, in arch formation, direct wave attack causes the rock layers to experience greater erosion at water level than above the water level. This process produces a breach in the narrow headland at the water level between the outermost part and the mainland. As the breach in the headland widens, the arch is formed.

Not all sea arches are curved. Due to the structural characteristics, some rocks produce arches that are nearly rectangular. The Tasman Arch on the island of Tasmania in Australia is a good example of such a feature. This huge rectangular arch was formed in distinctly jointed but resistant rock. The arch developed at and below sea level and rises to about 10 m above it. Its roof is almost at right angles to its sides because of the jointing. More commonly, arches are formed when wave erosion wears away the material near sea level and below, forming a curved arch because of the uniform nature of the rock.

Left: The London Bridge arch along the Great Ocean Road in southwestern Victoria, Australia, as it appeared in July 1986.

The lifetime of a sea arch is a limited one. The ongoing erosional processes that formed it will eventually destroy it. A good example of these forces at work took place at London Bridge on the western coast of Victoria in Australia. This limestone arch is only a few kilometers from the Twelve Apostles and had been there in apparently stable condition for as long as anyone knew. Then, in February, 1989, the arch suddenly collapsed, stranding two people on its seaward buttress; fortunately no one was on the arch itself. A helicopter plucked the stranded pair from the brand-new sea stack and brought them to safety.

The western coast of the United States also has a large number and variety of stacks and arches, formed in a range of rock types. Historical changes in several of these erosional features near La Jolla in southern California have been studied by the late Francis P. Shepard of the Scripps Institution of Oceanography and his colleague Gerry Kuhn. Using photographs supplemented by ship records and newspaper accounts, they pieced together data from more than a century of information on sea stacks and arches. An example is Cathedral Rock, a shore-connected sea arch with an upper surface that rose about 20 m above sea level. The earliest known photograph of this formation was taken some time before the 1870s; it

shows Cathedral Rock connected to the shore. The first dated photograph, in 1873, shows it separated from the mainland. The arch collapsed in 1906, leaving a large buttress at the seaward end. This stack continued to erode and completely disappeared in 1968.

The changes along rocky coasts are slow relative to one person's lifetime but are very rapid in terms of geologic time. Compared with coasts dominated by sand and other unlithified sediments, rocky coasts are relatively slow to respond to coastal processes. But like all features and all coasts, they inevitably change.

Right: The London Bridge arch after it collapsed, stranding two tourists in February 1989. The remains of the arch can be seen in the surf.

Epilogue

There is nothing more beautiful than the natural coast. So far, this book has discussed the principles that underlie its dynamic yet fragile nature. We have emphasized how even seemingly minor variations in the interactions among the controlling processes can bring about substantial changes in the coast; and how extreme events, such as hurricanes or tsunamis, can effect rapid and major changes. All of this discussion focused on the effects of natural processes on coastal formations. But what of the impact of human occupation on the coast? Here we will consider the effects of some of our attempts, both successful and unsuccessful, to control the coast.

For thousands of years, a large portion of the world's population has made their home along or near the coast. The original coastal settlers utilized natural settings without any significant modification, taking advantage of the various attributes of particular locations: In some cases, it was a natural harbor; in others, protection from adversaries or proximity to a desirable food supply. But eventually pressures such as increasing population, larger vessels, and industrialization resulted in substantial human modification of the coastal environment. These modifications

Rubble breakwater on the Mediterranean coast at Civitavecchia, Italy protecting small fishing boats.

generally were undertaken without regard for their impact on the natural coastal environment, and many have damaged the natural balances of the interaction between the local geomorphology and marine processes.

Human activity in the coastal environment is often directly or indirectly associated with construction of some kind and coastal construction, whether purposefully or not, affects the natural marine processes. Not only is the configuration of the coast changed, but so is its ecology. However, the principal focus of this book has been on the geologic forces and processes that affect the configuration of the coast, not its biology. In concluding, we continue this focus.

HARD COASTAL PROTECTION

Coastal construction that is intended to protect economic and recreational interests more often than not attempts to impede the natural erosional processes energized by weather, waves, and tides. Seawalls, breakwaters, groins, and jetties—all described as hard construction—serve this protective function in various ways, and the construction of these structures is called hardening the coast.

Seawalls

The landward movement of the shoreline is a normal and natural process. Limited sediment supply, rising sea level, or simply the washing over of a barrier during storms are some of the factors that cause shoreline retreat. These natural processes only become a problem when there are buildings or roads with economic value in the path of the moving shoreline.

Seawalls are vertical or sloping structures that are built along the shoreline in an attempt to stop erosion or at least retard it. They may be constructed of virtually any type of material—from plastic bags filled with sand to poured concrete armored with large boulders. Seawalls are among the most controversial coastal structures. Not only are they unsightly, but they are soon destroyed by wave scour and they reflect wave energy that may cause damage elsewhere.

To be successful, vertical, solid seawalls that are impermeable must be able to withstand the full impact of waves. Permeable structures, such as slotted walls or rip-rap (stone) walls constructed with large boulders, allow some wave energy to be absorbed through the structure; and their sloping walls can dissipate at least some wave energy. However, waves scour at the

Photograph showing the raising of the city of Galveston, Texas after the devastating hurricane of 1900. The houses on the right have been jacked up and filled underneath. The ones on the left are next in line for this procedure.

bases of all types of seawalls and eventually undermine them, thereby causing failure of the structures. As a result, seawalls provide only temporary protection.

An especially well designed and located seawall was built in Galveston, Texas, after an devastating hurricane in 1900 killed over 6000 people and destroyed much of the town. The wall was constructed in sections, with an upper elevation of 5.2 m above mean low water. The basic construction materials included wooden pilings, sheet-metal pilings, and rip-rap; the bulk of the wall, however, was a curved poured concrete surface. The seawall was impermeable, but rip-rap at its base and its curved face absorbed some wave energy. Furthermore, in a remarkable engineering and construction project, thousands of buildings, including several large stone churches, were raised an average of 2 m to the level of the crest of the wall. The seawall currently supports a six-lane highway and is the focal point for waterfront activities in the city. Although the Galveston seawall was severely damaged in 1915 during a hurricane having a storm surge of almost 4 m, there has been no major damage in the numerous storms since then. The wall is limited in its length, however; and the shoreline has retreated at the end of the seawall, where there is no stabilization, causing a large offset in the coast.

Breakwaters

Breakwaters are similar to seawalls, except that they are constructed seaward of the shoreline. They are called breakwaters because they are designed to reduce wave energy and to prevent the waves from eroding the shoreline. Because of the breakwaters' position in at least moderately deep water, they must be large and sturdy so that they can withstand the action of waves. They generally have vertical walls of poured concrete, but some also contain large rip-rap. Unlike seawalls, most breakwaters withstand the rigors of intense storms because they are in water deep enough to prevent scour under them.

These offshore structures may have a variety of configurations and locations, depending on the need. The simplest breakwater is a linear structure built parallel to the shore. This type of construction fairly widespread in Japan but is less common in the United States. Some breakwaters are attached to the shore; these are designed to form a protective harbor for moored vessels. This type of construction is fairly common in southern California and other coasts where natural harbors are absent.

Attaching a breakwater to the shore is essentially like building a dam along the shoreline. It interferes with the littoral transport system, by limiting or preventing sediment movement along the shoreline. As a consequence, large volumes of sediment accumulate updrift from the breakwater dam. The harbors associated with attached breakwaters often receive much of the accumulating sediment. To prevent the harbor from clogging up, bypassing systems consisting of a dredge and pump operation are permanently installed or harbors are dredged on a regular basis. In addition, severe erosion typically occurs downdrift of the breakwater, where sediment is prevented from accumulating.

Detached breakwaters that parallel the coast cause a "breakwater effect." By design, any breakwater limits the wave energy that reaches the shoreline. At either end of the breakwater, where there is no limiting influence, waves approach the shore without hinderance, producing longshore currents that carry sediment along the surf zone. The absence of wave energy landward of the breakwater causes an interruption of the littoral drift that produces a bulge in the shoreline behind the breakwater. Although accumulation of sediment is often one of the objectives of the breakwater construction, the trapping of sediment by the breakwater at one end may cause erosion at the other end, downdrift from the structure, because of the interruption of the littoral drift system. Further problems result if the shoreline progrades to the breakwater, filling in the open water

Long and high jetty trapping sand and protecting the harbor at St.-Malo, Ille et Viliaue Bretague, France. Tidal ranges here are among the highest in the world with high tide extending up to near the white paint. Note the people for scale.

between it and the mainland. If a series of breakwaters that are separated from each other is built, the opposite effect can occur, as wave action develops deep holes between the groins.

Groins

Groins are among the types of construction most frequently utilized in efforts to stabilize the coast. They are typically short structures that are attached perpendicular to the shoreline. Grouped into "groin fields," they extend across at least part of the beach and out into the surf zone. They are made of the same variety of materials used in seawalls.

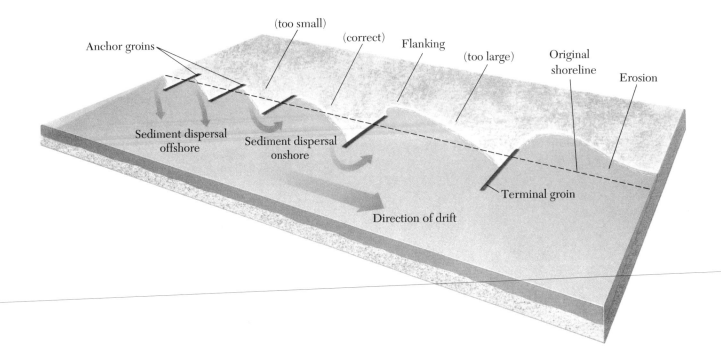

Anchor groins · (too small) · (correct) · Flanking · (too large) · Original shoreline · Erosion · Sediment dispersal offshore · Sediment dispersal onshore · Terminal groin · Direction of drift

Diagram showing how different spacing of groins causes different responses along the beach. If groins are too close, sediment is directed offshore and may be lost to the littoral drift system. When spacing is too great, there is erosion downdrift from each groin.

The purpose of groins is to trap sediment that moves through the littoral drift system, thereby maintaining beach in areas that would otherwise experience erosion. Given the proper length, elevation, and spacing, groins will permit sediment to accumulate until the groin field is buried. Sediment will then bypass the buried groin without causing significant downdrift erosion, with the exception of some temporary erosion caused by major storms. In reality, however, the typical result of groin installation is a smaller-scale version of the updrift accumulation and downdrift erosion that occurs with attached breakwaters. In some cases, erosion is not impeded significantly, and the groins become detached from the beach and have no positive effect at all.

The best examples of successful groins are found along the North Sea coasts of the Netherlands and Germany, where the coasts have been actively protected for over a thousand years. Here the severe winter storms and the strong littoral drift require large groins—hundreds of meters long and several meters high. Many have become buried and there is little asymmetry in the sediment accumulation on either side of the individual groins, a testimony to their success. On this coast, groins have been in place for nearly a century without major maintenance or reconstruction.

Jetties

Jetties are much like groins except that they are typically larger and are located at tidal inlets. Jetties are constructed to stabilize one or both sides of an inlet to permit its continued use for navigation. In some cases, the inlet is deepened after jetty construction to accommodate the passage of deep-draft vessels.

Like groins, jetties interrupt the littoral drift system. Sediment accumulates on the updrift side and is prevented from reaching the downdrift side of the inlet, thus inducing erosion. Because jetties can be quite long, in some cases a kilometer or more, a large amount of sediment can accumulate on the updrift side. Occasionally the accumulation becomes so large that it extends past the seaward end of the jetty and into the inlet. However, systems to allow sediment to bypass jetties are becoming a part of the maintenance programs for many jettied inlets.

A good example of jetty-induced erosion is found near Ocean City, Maryland. In 1933, an inlet was cut south of the city by a hurricane. Jetties were constructed shortly thereafter to protect the inlet from a littoral drift that transported sediment at a rate of about 140,000 m^3 per year. No bypass system was installed, however. As a result, the shoreline on Assateague Island, the downdrift portion of the system, has retreated about a kilometer.

SOFT COASTAL PROTECTION

Within the past few decades there has been a growing trend away from hard construction toward so-called soft means of coastal protection. These methods include beach nourishment and dune stabilization. This approach avoids the use of materials foreign to the coastal environment that is being protected and thus does not incorporate any of the traditional types of hard construction. The natural materials used are chosen for their aesthetic consistency and their compatibility with natural coastal processes.

Beach Nourishment

One of the earliest approaches to soft construction was the replacement of sand on eroding beaches. Although beach nourishment has been accomplished in isolated locations since the turn of the century, this method has been practiced systematically and on a widespread basis only since the 1970s. In the earliest nourishment efforts, an eroding beach was resupplied

Pumping sand onto the beach at Wrightsville, North Carolina to nourish a chronically eroding coast. The sand is typically pumped from a dredge in slurries of 10 to 15 percent sediment.

with the closest available sand, taken from an adjacent dune or obtained by dredging close offshore or in a shoaling inlet nearby. Little attention was paid to the texture of the nourishment material or to the design of the nourished area. As a result, the nourishment sand was often quickly washed away.

More recently the size and shape of the beach have been designed to be compatible with the adjacent depth and wave conditions. Replacements materials are selected to be as similar to the original beach material as possible and are taken from sites, called borrow areas, that will be minimally damaged by their removal.

Some beach nourishment projects have been both successful and cost effective. For several years in Miami Beach, Florida, there was essentially no beach in front of the many luxury hotels. A multiyear project to replenish over 15 km of beach with millions of cubic meters of sand—at a cost of $65 million—was completed in 1980. The sand was extracted by a suction dredge from a shallow offshore area and pumped onshore. It was then carefully placed by earthmovers and bulldozers to conform to predetermined design criteria. With stabilizing vegetation now present on the backbeach area, the beach is today almost as wide as it was immediately after the project's completion.

Smaller projects on Captiva Island and on Sand Key on the Gulf Coast of Florida have been similarly successful. However, the benefits of other nourishment projects have not lasted very long. At some locations, erosion removed nearly all of the sand placed on the beach within a couple of years, even when the design called for a lifetime of about 10 years. An example was the nourishment of the beaches south of Cape Canaveral on the eastern coast of Florida, where the emplaced sand was too fine grained and most was eroded away in less than two years. Another example was the unanticipated removal of much nourishment material at Myrtle Beach, South Carolina, by Hurricane Hugo in 1989, an event that could not have been foreseen. Despite the problems, beach nourishment has almost totally replaced seawall and other types of hard construction, both because of legislation and because this method is cost effective.

Dune Stabilization

Coastal sand dunes have long been recognized as critical to the protection of the shoreline, but a look at a well-worn footpath or a blowout demonstrates dramatically how easily sand on an unvegetated dune can be eroded. Wind activity causes migration of sand from dunes and wave attack erodes their seaward sides. Dunes can be stabilized to reduce erosion and prevent their landward movement or seaward destruction. Their maintainance and growth has become a primary objective of coastal management.

The current dune management effort is twofold: to build new dunes and to protect existing dunes from destruction. Various types of fences have been used for many years to trap windblown sand. The simplest configuration consists of a single fence placed at the back of the beach, in front of the existing dune, although other, more complicated arrangements are possible. If the fence entraps sufficient sand, the dune grows and is at least partially held in place by the fencing. The fence may be buried and a second fence installed to further increase the dune height. One difficulty is that the anchoring provided by the fences prevents any dune migration during wave washover, so the result may be that the erosion and removal of the entrapped sediment by wave attack is accelerated.

In concert with fencing, vegetation has proved to be both an excellent dune stabilizer and dune builder. Root networks hold sediment in place, and the leaves and grass blades baffle winds to trap sediment and hold it in place as well. Walkovers and footpaths protect vegetation from destruction and protect the dune surface from wind erosion, even in the absence of plants. Paths are oriented at an angle to the dominant wind direction to prevent erosion.

There are other effective and inexpensive approaches to dune stabilization that are variations on the fencing and vegetation methods. Along the North Sea coast of the Netherlands, small fencelike structures of shrub twigs and branches are placed in shallow ditches dug at the bases of and

Planting of dune grass to aid in the reconstruction of dunes in the shadow of erosion remnants of older dunes near St. Helens on the east coast of Tasmania, Australia.

throughout the foredunes. Retired coastal residents provide the bulk of the laborforce, and for their work they receive a supplement to their pensions. The short fences are quite effective in trapping sand and also in stabilizing existing dunes. Along the Texas Gulf Coast, residents use their discarded Christmas trees to trap sand in front of the existing dunes. This approach is almost without cost, it is effective, and the trees biodegrade.

This series of fences in Wales shows the effects of this type of structure on dune development. The fence not only helps to accumulate a dune along its extent, it also results in the accumulation of a secondary dune landward of it.

Dredge and Fill Construction

Although certain coastal structures are specifically intended to intervene in the natural coastal processes, many serve other purposes. Nevertheless they also alter these processes. For example, causeways and buildings for residential and industrial use on reclaimed coastal land rely on dredging and fill operations—both environment-altering operations.

Causeways

The easiest and least expensive way to create a roadbed is to fill in a shallow coastal aquatic environment creating a causeway. This otherwise solid construction is interrupted by bridges for boat traffic and short overpasses for water circulation. Additional bridges are built when the water is too deep to be filled. A causeway acts much like a dam and greatly inhibits tidal flushing, creating two major problems for the affected bodies of water: increased pollution and inlet migration or closure.

Most coastal waters receive considerable chemical pollution from the adjacent populated areas. Because the tidal circulation is restricted pollutants are not removed from enclosed bodies of water and the water quality deteriorates.

Restricted circulation also affects the tidal prism and inlets served by tidal flow are stabilized in an equilibrium condition that depends in part on tidal prism. By decreasing this prism, causeways upset this equilibrium condition. The consequence is a decrease in inlet size, migration of the inlet, and, in extreme cases, closure of the inlet. Dunedin Pass on the Gulf Coast of Florida deteriorated from a width of over 150 m and a maximum depth of 6 m to eventual closure over a 60-year period. The problem began with the completion of the Clearwater Causeway south of the inlet in 1922. After another causeway was constructed in 1964 on the other side of the inlet about 15 km north of the first causeway, the rate of inlet instability accelerated.

Residential and Industrial Buildings

Waterfront property along coastal bays is scarce and expensive. Therefore, for many years sediment was dredged from intertidal zones or shallow bays to create upland areas for residential or industrial development.

Residential areas constructed on filled land often include narrow finger canals between the filled areas. These canals are generally disastrous for the environment. Because virtually all of the canals are dead-ended, the tides cannot circulate throughout the canals and poor water quality develops. The decline in water quality is compounded by a high rate of pollution influx from lawns, gardens, storm-water runoff, and other products of human occupation.

Like causeway construction, dredging and filling reduces the area of the body of water and therefore reduces the tidal prism for the adjacent inlets, which become unstable and even close unless additional structures are built to maintain the tidal flow.

The coast is dynamic, complex, and fragile. In building along it, we have interfered with the natural processes of littoral drift and tidal flow. But as more is understood about how coastal processes interact with various geological environments and changes in sea level, we are better able to shape our activities to harmonize with the natural coastal dynamics.

Suggested Readings

General

Barnes, R. S. K., ed., 1977. *The Coastline*. London: John Wiley & Sons, Inc. Covers all of the coastal environments.

Bird, E. C. F., 1985. *Coastline Changes: A Global Review*. New York: John Wiley & Sons, Inc. A brief survey of all coastal countries in the world but lacking in quality illustrations.

Bird, E. C. F., and M. L. Schwartz, eds., 1985. *The World's Coastlines*. New York: Van Nostrand Reinhold. A systematic survey of the coasts of the world that stresses beaches.

Carter, R. W. F., 1988. *Coastal Environments*. New York: Academic Press. A good overview of all coastal types with a British Isles flavor.

Davies, J. L., 1980. *Geographic Variation in Coastal Development*. New York: Longman Group, Ltd. A nice global picture of coastal characteristics.

Davis, R. A., 1985. *Coastal Sedimentary Environments*. New York: Springer-Verlag New York Inc. Contains comprehensive chapters on various coastal environments by experts.

Shepard, F. P., and H. R. Wanless, 1971. *Our Changing Coastlines*, New York: McGraw-Hill Inc. A tour of the United States coastline via maps and aerial photos showing how changes have taken place during historical time.

Chapter 1: **Plate Tectonics and the Coast**

Bolt, B. A., 1993. *Earthquakes and Geological Discovery*. New York: W. H. Freeman and Company Publishers. A well-presented general treatment of the dynamics of the earth's crust.

Stanley, S. M., 1989. *Earth and Life Through Time*, 2/e. New York: W. H. Freeman and Company Publishers. Discussion of physical and biological evolution of the earth's crust.

Chapter 2: **The Changing Sea Level**

Eisma, D., ed., 1995. *Climate Change: Impact on Coastal Habitation*. Boca Raton, Fla.: CRC Press. Very interesting contributions on the subject of how climate change, as expressed in sea level rise, will change our occupation of the coast.

National Research Council, 1987. *Responding to Changes in Sea Level: Engineering Implications*. Washington, D.C.: National Academy of Science Press. Evaluation of all types of scenarios by a blue-ribbon committee.

Titus, J. G., and V. K. Narayanan, 1995. *The Probability of Sea Level Rise*. Washington, D.C.: U. S. Environmental Protection Agency. A statistical look at a wealth of environmental data to help predict sea level rise over the next couple of centuries.

Wind, H. G., ed., 1987. *Impact of Sea Level Rise on Society*. Rotterdam: A. A. Balkema.

Chapter 3: **Processes that Shape the Coast**
Bowditch, N., 1966. *American Practical Navigator*. Washington, D.C.: U. S. Naval Hydrographic Office. Basics of hydrodynamic processes.

Campion, D., 1989. *The Book of Waves*. Santa Barbara, Calif.: Arpel Graphics. A beautifully illustrated book that describes the basics of waves.

Defant, A., 1958. *Ebb and Flow: The Tides of the Earth*. Ann Arbor, Mich.: University of Michigan Press. A comprehensive guide to tides and tidal processes.

Komar, P. D., ed., 1983. *Handbook of Coastal Erosion and Processes*. Boca Raton, Fla.: CRC Press Inc. Discussions of coastal processes and environments by experts.

Chapter 4: **Estuaries, Salt Marshes, and Tidal Flats**
Castanares, A., and F. B. Phleger, eds., 1969. *Coastal Lagoons—A Symposium*. Mexico City: UNAM-UNESCO. A collection of important papers on coastal lagoons with emphasis on the Gulf of Mexico.

Cronin, L. E., ed., 1975. *Estuarine Research*, vol. 2. New York: Academic Press. An excellent collection of papers on inlet and estuarine dynamics and sedimentology.

Dyer, K. R., 1986. *Coastal and Estuarine Sediment Transport*. New York: Wiley-Interscience.

Perillo, G. M. E., ed., 1995. *Geomorphology and Sedimentology of Estuaries*, Dev. in Sedimentology No. 53. Amsterdam: Elsevier Science. An excellent treatment of all geologic aspects of estuaries.

Chapter 5: **Deltas—Where Rivers Unload Their Deposits**
Broussard, M. L., ed., 1975. *Deltas—Models for Exploration*. Houston, Tex.: Houston Geological Society. Details on many of the world's best known deltas.

Coleman, J. M., 1981. *Deltas, Processes of Deposition and Models for Exploration*. Minneapolis, Minn.: Burgess Publishing, Inc. A short volume on the morphology and stratigraphy of deltas; lacks detail concerning processes.

Morgan, J. P., ed., 1975. *Deltaic Sedimentation*, SEPM Spec. Publ. No. 15. Tulsa, Okla. Sedimentology and stratigraphy of several of the world's important ancient and modern deltas.

Oti, M. and G. Postma, eds., 1995. *The Geology of Deltas*. Rotterdam: A. A. Balkema Publishers.

Chapter 6: **Beaches, Dunes, and Barriers**
Davis, R. A., ed., 1994. *Geology of Holocene Barrier Island Systems*. Heidelberg: Springer-Verlag New York, Inc. The most comprehensive volume on barrier islands; organized by geographic regions.

Greenwood, B., and R. A. Davis, eds., 1984. *Hydrodynamics and Sedimentation in Wave-Dominated Coastal Environments*, Dev. in Sedimentology No. 39. Amsterdam: Elsevier Science. Coverage of beach and nearshore environments throughout the world.

King. C. A. M., 1972. *Beaches and Coasts*, London: Edward Arnold. Comprehensive but dated coverage of beaches with emphasis on the United Kingdom.

Leatherman, S. P., ed., 1979. *Barrier Islands*. New York: Academic Press. Compilation of several good articles on specific barrier island systems.

Nordstrom, K. F., N. Psuty, and B. Carter, eds., 1991. *Coastal Dunes: Form and Processes*. New York: John Wiley & Sons Inc. Details on coastal dunes all over the world.

Chapter 7: **Rocky Coasts**
Trenhaile, A. S., 1987. *The Geomorphology of Rock Coasts*, Oxford Studies in Geography. Oxford: Clarendon Press. The most comprehensive volume on rocky coasts; includes numerous examples.

Epilogue
Coastal Engineering Research Center, 1984. *Shore Protection Manual*. Vicksburg, MS: U. S. Army, Corps of Engineers. The manual for coastal engineering practice. It includes chapters on waves and on beaches.

Pilkey, O. H., and K. Dixon, 1996. *The Corps and the Shore*. El Centro, Calif.: Island Press. This book takes issue with many of the coastal activities of the Corps on Engineers.

Pilkey, O. H., ed. *Living with the _____ Shore*. Durham, N.C.: Duke University Press. A series of books on coastal states that addresses the problems of living along the open coast.

Sources of Illustrations

Cartography by Susan Johnston Carlson; geological cross-sections by George V. Kelvin; and line illustrations by Precision Graphics

Cover image
AUSCAPE International

Page 1
Charles Gurche

Page 2
James Valentine

Page 4
National Archives

Page 6
Baron Wolman

Page 8
Bildarchiv Preussischer Kulturbesitz, Berlin

Page 9
Modified from R. S. Dietz and J. C. Holden, 1970, "Reconstruction of Pangea: breakup and dispersion of continents, permian to present," *Journal of Geophysical Research*, v. 75, p. 4939–4956; data from A. L. DuToit, 1937, *Our Wandering Continents*, Oliver and Boyd, Edinburgh

Page 11
From A. Wegener, 1924 (reprinted 1955), *The Origins of Continents and Oceans*, Dover, Mineola, New York

Page 12
Adapted from J. Francheteau, "The oceanic crust," © 1983

by *Scientific American, Inc.*; plate motions modified from the work of J. B. Minster and T. H. Jordan

Page 13
Infermer

Page 14
From R. A. Davis, Jr., 1991, *Oceanography*, second edition, Wm. C. Brown, Dubuque, Iowa

Page 15
From map by W. C. Pitman, III, R. L. Larson and E. M. Herron, 1974, Geological Society of America

Page 17
From Baranzangi and Dorman, 1969, *Bulletin of Seismological Society of America*

Page 19
From R. A. Davis, Jr., 1991, *Oceanography*, second edition, Wm. C. Brown, Dubuque, Iowa

Page 20
From W. Hamilton, U. S. Geological Survey

Page 21
From D. L. Inman and C. F. Nordstrom, 1971, "On the tectonic classification of coasts," *Journal of Geology*

Page 22
From D. L. Inman and C. F. Nordstrom, 1971, "On the tectonic classification of coasts," *Journal of Geology*

Page 23
Loren McIntyre

Page 24
Adapted from P. D. Komar, 1976, *Beach Processes and Sedimentation*, Prentice-Hall, Inc., Englewood Cliffs, New Jersey

Page 25
Robert Dill, U. S. Navy

Page 26
Loren McIntyre

Page 27
Adapted from R. Siever, 1988, *Sand*, Scientific American Library, New York

Page 28
Peter Kresan

Page 29
GEOPIC, Earth Satellite Corporation

Page 30
Jim Brandenburg/Minden Pictures

Page 32
Nikita Ovsyanikov/Planet Earth Pictures

Page 33
Loren McIntyre

Page 35
GEOPIC, Earth Satellite Corporation

Page 38
Peter Kresan

Page 40
Loren McIntyre

Page 41
R. A. M. Schmidt

Page 43
From P. D. Komar and D. B. Enfield, 1987, "Short-term sea-level changes and coastal erosion," in D. Nummedal, O. H. Pilkey and J. D. Howard, eds., *Sea Level Change and Holocene Coastal Development*, SEPM Special Publication No. 41, Tulsa, Oklahoma, p. 17–27

Page 44
From P. D. Komar and D. B. Enfield, 1987, "Short-term sea-level changes and coastal erosion," in D. Nummedal, O. H. Pilkey and J. D. Howard, eds., *Sea Level Change and Holocene Coastal Development*, SEPM Special Publication No. 41, Tulsa, Oklahoma, p. 17–27

Page 45
Modified from P. D. Komar and D. B. Enfield, 1987, "Short-term sea-level changes and coastal erosion," in D. Nummedal, O. H. Pilkey and J. D. Howard, eds., *Sea Level Change and Holocene Coastal Development*, SEPM Special Publication No. 41, Tulsa, Oklahoma, p. 17–27

Page 47
Richard A. Davis, Jr.

Page 48
From A. L. Bloom, 1978, *Geomorphology*, Prentice-Hall, Inc., Englewood Cliffs, New Jersey; data from G. deQ. Robin, 1964, *Endeavour*, v. 23, p. 102–107

Page 49
Loren McIntyre

Page 51
From M. Ince, ed., 1990, *The Rising Seas*, Proceedings of the Cities on Water Conference, Venice, Italy

Page 52
From the United Nations Environment Program, 1989

Page 53
From A. Cox, 1968, "Polar wandering, continental drift, and the onset of Quaternary glaciation," in *Causes of Climatic Change* (INQUA, 7th Congress, Proceedings), Meteorological Monograph, v. 8, p. 112–125, in J. T. Andrews, 1975, *Glacial Systems*, North Scituate, Mass., Duxbury Publishing Co.

Page 56
Modified from J. T. Andrews, 1975; based on data from R. F. Flint, 1971

Page 58
National Air Photo Library of Canada

Page 59
J. Curray, 1969

Page 61
Adapted from S. D. Hicks, 1972, *Shore and Beach*, v. 40, p. 20

Page 62
From 1987, *Responding to Changes in Sea Level, National Academy of Sciences*, National Academy Press, 148 p.; data from J. C. Stevenson, L. G. Ward, and M. S. Kearney, 1986, "Vertical accretion in marshes with varying rates of sea level rise," in D. Wolf, ed., *Estuarine Variability*, Academic Press, New York, p. 241–260

Page 63
Modified from V. Gornitz and S. Lebedeff, 1987, "Global sea-level changes during the past century," in D. Nummedal, O. H. Pilkey and J. D. Howard, eds., *Sea Level Rise and Coastal Development*, SEPM Special Publication No. 41, Tulsa, Oklahoma, p. 3–16

Page 65
From M. Ince, ed., 1990, *The Rising Seas*, Proceedings of the Cities on Water Conference, Venice, Italy

Page 66
Warren Bolster/Tony Stone Images

Page 70
top, from W. Bascom, "Ocean waves," © 1959 by *Scientific American, Inc.*; bottom, Tony Stone Images

Page 72
Warren Bolster/Tony Stone Images

Page 76
Richard Johnson/Tony Stony Images

Page 79
bottom, Japan Meteorological Agency

Page 85
Modified from P. D. Komar, 1976, *Beach Processes and Sedimentation*, Prentice-Hall, Inc., Englewood Cliffs, New Jersey

Page 95
Dana Fisher/San Diego Union

Page 96
From W. T. Fox and R. A. Davis, 1976, "Weather patterns and coastal processes," in R. A. Davis and R. L. Ethington, eds., *Beach and Nearshore Sedimentation*, SEPM Special Publication No. 24, Tulsa, Oklahoma, p. 1–23

Page 97
Library of Congress

Page 99
European Satellite Agency

Page 100
Murray & Associates/Tony Stone Images

Page 102
Tupper Ansel Blake

Page 103
Richard A. Davis, Jr.

Page 104
Loren McIntyre

Page 105
From M. M. Nichols and R. B. Biggs, 1985, "Estuaries," in R. A. Davis, ed., *Coastal Sedimentary Environments*, Springer-Verlag, Inc., New York, p. 77–186

Page 107
Modified from H. Postma, 1980, "Sediment transport and sedimentation," in E. Olausson and I. Cato, eds., *Chemistry and Biogeochemistry of Estuaries*, John Wiley and Sons, Chichester, p. 153–186

Page 110
From R. W. Dalrymple, C. L. Amos and S. B. McCann, 1982, "Beach and nearshore depositional environments of the Bay of Fundy and southern Gulf of St. Lawrence," in *Fieldtrip Guidebook 6A*, International Association of Sedimentologists, Hamilton, Ontario, p. 15

Page 111
From H. Postma, 1961, "Transport and accumulation of suspended matter in the Dutch Wadden Sea," *Netherlands Journal of Sea Research*, v. 1, p. 148–190

Page 112
Ric Ergenbright

Page 113
Richard A. Davis, Jr.

Page 115
Patrick Lorne, Explorer Archives

Page 116
After H. Postma, 1967, "Sediment transport and sedimentation in the marine environment," in G. H. Lauff, ed., *Estuaries*, American Association for the Advancement of Science, Special Publication 83, Washington, D. C., p. 158–179 and L. M. J. U. Van Straaten, and P. H. Kuenen, 1957, "Accumulation of fine grained sediments in the Dutch Wadden Sea," *Geo. en. Mijnbouw (NW Ser.)*, v. 19, p. 329–354

Page 118
top, from G. D. Klein, 1977, *Clastic Tidal Facies*, Continuing Education Publishing Co., Champaign, Ill., 149 p.; after H. -E. Reineck and F. Wunderlich, "Seitmessungen und Beseitenschickten," *Natur und Museum*, v. 97, p. 193–197; bottom, Richard A. Davis, Jr.

Page 119
Joost Terwindt

Page 120
Richard A. Davis, Jr.

Page 122
James Valentine

Page 124
James Valentine

Page 125
James Valentine

Page 127
Richard A. Davis, Jr.

Page 131
Richard A. Davis, Jr.

Page 132
GEOPIC, Earth Satellite Corporation

Page 135
Data from L. D. Wright, 1982, "Deltas," in M. L. Schwartz, ed., *Encyclopedia of Beaches and Coastal Environments*,

Hutchinson Ross, Stroudsburg, Pennsylvania, p. 358–368 and J. D. Milliman and R. H. Meade, 1983, "World-wide delivery of river sediment to the oceans," *Journal of Geology*, v. 91, p. 1–21

Page 136
EOSAT/Explorer Archives

Page 137
From C. R. Kolb and J. R. Van Lopik, 1966, "Depositional environments of Mississippi River deltaic plain, southeastern Louisiana," in M. L. Shirley, ed., *Deltas in Their Geological Framework*, Houston, Geological Society, Houston, Texas, p. 17–61

Page 138
Miles O. Hayes

Page 139
Richard A. Davis, Jr.

Page 140
EROS/Mobile Exploration and Producing Technical Center

Page 141
Modified from Coleman and Gagliano, 1964

Page 143
From W. E. Galloway, 1975, "Process framework for describing the morphologic and stratigraphic evolution of deltaic depositional systems," in M. L. Broussard, ed., *Deltas: Models for Exploration*, Houston Geological Society, Houston, Texas, p. 87–98

Page 145
From L. D. Wright and J. M. Coleman, 1973, "Variations in morphology of major river deltas as functions of ocean wave and river discharge regimes," *American Association of Petroleum Geologists Bulletin*, v. 57, p. 370–398

Page 146
GEOPIC, Earth Satellite Corporation

Page 147
left and right, from L. D. Wright and J. M. Coleman, 1973, "Variations in morphology of major river deltas as functions of ocean wave and river discharge regimes," *American Association of Petroleum Geologists Bulletin*, v. 57, p. 370–398

SOURCES OF ILLUSTRATIONS

Page 148
top left, from G. P. Allen, D. Laurier, and J. Thouvenin, 1979, *Notes et Memoires No. 15*, Total, Paris, Compagnie Francaise des Petroles, 156 p.; top right, from M. L. Shirley, ed., 1966, *Deltas in Their Geologic Framework*, Houston Geological Society, Houston, Texas; bottom left, from J. R. L. Allen, 1970, "Sediments of the modern Niger Delta: a summary and review," in J. R. Morgan, and R. H. Shaver, eds., *Deltaic Sedimentation*, SEPM Special Publication No. 15, Tulsa, Oklahoma, p. 138–151

Page 150
top, Loren McIntyre; bottom, Historic New Orleans Collection, Museum/Research Center

Page 151
Peter Kresan

Page 152
James Valentine

Page 156
Ric Ergenbright

Page 157
Larry Ulrich

Page 159
Steve Rose/Rainbow

Page 161
James Valentine

Page 162
James Valentine

Page 163
Larry Ulrich

Page 164
Reg Morrison/AUSCAPE International

Page 165
SKAGENS MUSEUM, The Skaw Museum, Denmark

Page 167
Modified from H. Blatt, G. V. Middleton and R. Murray, 1980, *Origin of Sedimentary Rocks*, second edition, Prentice-Hall, Inc., Englewood Cliffs, New Jersey

Page 168
William T. Fox

Page 170
From A. J. Scott, R. A. Hoover and J. H. McGowen, 1969, "Effects of Hurricane Beulah, 1967, on Texas coastal lagoons and barriers," *Members of the International Symposium on Coastal Lagoons*, UNAM-UNESCO, p. 221–236

Page 173
Albert C. Hine

Page 176
Bill Jordan

Page 179
Richard A. Davis, Jr.

Page 181
From M. O. Hayes and T. W. Kana, eds., 1976, "Terrigenous Clastic Depositional Environments." *Technical Report No. 11-CRC*, Coastal Research Division, University of South Carolina, Columbia, S.C., 302 p.

Page 183
Richard A. Davis, Jr.

Page 184
Carr Clifton

Page 186
Art Wolfe

Page 187
Rank Siteman/Rainbow

Page 189
top, from R. A. Bagnold, 1939, "Interim report on wave pressure research," *Journal of the Institution of Civil Engineers*, London, v. 12, p. 202–226; bottom, from R. A. Bagnold, 1939, "Interim report on wave pressure research," *Journal of the Institution of Civil Engineers*, London, v. 12, p. 202–226

Page 192
Rex Elliot

Page 193
Travis Amos

Page 194
David Phillips/Planet Earth Pictures

Page 196
Dennis Harding/AUSCAPE International

Page 197
Carr Clifton

Page 198
From E. C. F. Bird, 1976, *Coasts, an Introduction to Systematic Geomorphology*, Australian National University Press, Canberra, p. 282

Page 199
Larry Ulrich

Page 202
Richard A. Davis, Jr.

Page 203
Richard A. Davis, Jr.

Page 204
Art Wolfe

Page 207
Rosenberg Library, Galveston, Texas

Page 209
Philip Plisson/Explorer

Page 210
From R. W. G. Carter, 1988, *Coastal Environments*, Academic Press, London, 617 p.

Page 212
Jack Dermid/Photo Researchers

Page 214
Reg Morrison/AUSCAPE International

Page 215
David Woodfall/Natural History Photo Agency

Index

Diffraction, wave, 77–78
Dissipative beach, 155
Distributary, 133
Distributary mouth bar, 141–142
Drainage divide, 24, 33–34
Dredge and fill construction, 215–216
Dredging
    for beach nourishment, 212
    for breakwater construction, 208
Drumstick barrier islands, 181–182
Dune(s), 160–166
    blowouts of, 165
    formation and distribution of, 160–164
    inland, 163–164
    migration of, 164–166
    stabilization of, 213–215
Dune ridge, 162–163

Earthquakes
    distribution of, 17
    sea level changes and, 40–42, 49
    tsunamis and, 69, 78–81
Ebb tide, 85
Ebb tide delta, 172, 174–175
    erosion of, 177
El Niño, sea level changes and, 44–46
Emiliani, Cesare, 55
End moraines, 187
Eolianite deposits, on rocky coasts, 188, 192–193
Equatorial tides, 88
Erosion
    of deltas, 142
    differential, 201
    at leading edge coasts, 25
    of rocky coasts, 189–193
    wave action and, 189–190
Erosive beach, 158–159
Estuaries, 101–131
    agglomerates in, 111

aggregates in, 111
Bay of Fundy as, 129–131
biogenic material in, 113
boundary layer of, 110
characteristic features of, 101, 103
Chesapeake Bay as, 128–129
circulation in, 104–108
definition of, 103
vs. deltas, 133–134
flocculation in, 111–112
formation of, 33
fully mixed, 105
head of, 103
vs. lagoons, 101
mouth of, 103
partially mixed, 104–105
river-dominated, 109–110
salinity in, 106, 171
San Antonio Bay as, 126–128
seawater-freshwater interaction in, 104–108
sediment deposition in, 108–124
    in mangrove swamps, 122, 123, 125–126
    in salt marshes, 121–124
    in tidal flats, 115–121
stratified, 104
tidal bore of, 104
tidal flats of, 115–121
tidal range of, 103–104
tide-dominated, 110–114
time-velocity curve for, 110–111
Willapa Bay as, 129
Estuarine organisms
in Bay of Fundy, 131
bioturbation by, 120–121
in Chesapeake Bay, 128–129
in San Antonio Bay, 127–128
sediment deposition by, 114
in Willapa Bay, 129
Evaporation, rocky coast erosion and, 190
Ewing, Maurice, 13

Fences, for dune stabilization, 213–215
Fetch, 69
Filter feeders
    bioturbation by, 120–121
    estuarial sediment deposition by, 114
Finger canals, 216
First-order features, 22, 31
Fish. See Aquatic animals
Flocculation, in estuaries, 111–112
Flooding, sea level rise and, 64–65
Flood tide, 85
Flood tide delta, 172, 173–174
Fluid withdrawal, sea level changes and, 47–48
Foredune ridge, 162–163
Foreshore, 154
Freezing, rocky coast erosion and, 190
Fresh water, in estuaries, 104–108
Froude number, 111

Galloway, William, 143
Ganges-Brahmaputra delta, as tide-dominated delta, 145
Geomorphologists, 3
    military, 4–5
Glacial drift, cliffed coast formation and, 187
Glaciation. See also Ice sheets
Glaciation periods
    oxygen isotope dating of, 54–55
    Pleistocene, 53–57
Global warming, sea level changes and, 51–52, 61–63
Good sorting, of beach components, 156
Grasses
    dune formation and, 162
    for dune stabilization, 214
    in salt marsh, 121–124
Gravel beaches, 156
Gravity, tides and, 85–88

Ocean floor, formation of, 15–17, 18–19
Oceanic crust. *See also* Lithosphere
  dating of, 14–15, 46–47
  density of, 15
Oceanic islands, dating of, 14
Oceanic ridge system, 13
  earthquakes in, 17
  rift valleys of, 18
Oceanic trenches
  earthquakes and, 17
  formation of, 18
Ocean temperature, sea level
    changes and, 42–46
Office of Naval Research, 5
Oxygen isotope techniques, for
    glaciation dating, 54–55

Pangaea, 9
Panthalassa, 9
Perigee tides, 90
Peru Current, El Niño and, 44–46
Plane bed, in tide-dominated
    estuary, 110
Plants
  dune formation and, 162
  for dune stabilization, 214
  rocky coast erosion and, 191
  salt marsh, 123
Plate boundaries, 10–21
Plate movement, 18–19
  nature of, 18–19
  rate of, 19
Plate tectonics, 18–37
  coast classification and, 20–37
  continental drift theory and, 8–18, 20
  definition of, 20
  delta formation and, 134–135
  first-order coastal features and, 22, 31, 34
  plate movement and, 18–20
  sea level changes and, 40–42, 49, 50

Platforms
  abrasion, 160
  shore, 200
  subtidal, 200
Pleistocene glaciation, 53–57
Plunging breaker, 74
Pneumatophore, 125–126
Polar ice caps. *See* Ice sheets
Positive storm tide, 95
Postma, H., 116
Pritchard, Donald, 106
Progradation, barrier island, 183
Pruitt, Evelyn, 5
Pull-apart basins, 27

Quarternary Period, glaciation in, 52–57

Rebound, lithospheric, 57–58
Red beds, 130
Red mangrove, 125
Reeds (*Phragmites*), 123
Reflection, wave, 76–77
Reflective beaches, 154, 158, 160
Refraction, wave, 78
Regional subsidence, sea level
    changes and, 46–48
Residential construction, dredge and
    fill, 216–217
Ridge and runnel, 154
Rift valleys, 18
Rip channel, 158
Rip currents, 84, 157–158
Ripples, 69
  in tide-dominated estuaries, 110
Rip-rap walls, 206
River-dominated deltas, 144
River-dominated estuaries, 109–110
Rock
  chemical weathering of, 192–193
  erosion patterns in, 196
Rocky coasts, 185–203
  beaches on, 198
  chemical weathering of, 192–193

cliffed, 186–188, 195–198
  erosional remnants of, 200–203
  erosion of
    biological causes of, 191
    chemical causes of, 191–193
    differential, 200–201
    physical causes of, 188–191
    sites of, 193–194
  geology of, 185–188
  geomorphology of, 194–203
  glacial-drift, 187
  horizontal features of, 199–200
  origins of, 185–188
  shore platforms of, 199–200
  subtidal platforms of, 200
  terraces of, 200
  tidal range on, 190
  types of, 186–188
  vertical features of, 195–198
  wave action at, 188–190
Rogue wave, 72
Runnel, 154

Salinity
  in estuaries, 106–108, 171
  in lagoons, 171
Salt marshes, 121–124
San Andreas fault system, 26
San Antonio Bay, as estuary, 126–128
Sand. *See also under* Sediment
  beach, 156
Sandbars
  distributary mouth, 141–142
  formation of, 154–155
Sand dunes. *See* Dune(s)
Sand replenishment, 213–215
Sand transport
  longshore currents and, 83
  in tide-dominated estuaries, 110–114
Sand waves, 112
São Francisco delta, as wave-
    dominated delta, 146–147